D0895464

Analog VLSI Circuits for the Perception of Visual Motion

Analog VLSI Circuits for the Perception of Visual Motion

Alan A. Stocker
*Howard Hughes Medical Institute and Center for Neural Science,
New York University, USA*

John Wiley & Sons, Ltd

Copyright © 2006   John Wiley & Sons Ltd, The Atrium, Southern Gate, Chichester,
West Sussex PO19 8SQ, England

Telephone (+44) 1243 779777

Email (for orders and customer service enquiries): cs-books@wiley.co.uk
Visit our Home Page on www.wiley.com

All Rights Reserved. No part of this publication may be reproduced, stored in a retrieval system or transmitted
in any form or by any means, electronic, mechanical, photocopying, recording, scanning or otherwise, except
under the terms of the Copyright, Designs and Patents Act 1988 or under the terms of a licence issued by the
Copyright Licensing Agency Ltd, 90 Tottenham Court Road, London W1T 4LP, UK, without the permission in
writing of the Publisher. Requests to the Publisher should be addressed to the Permissions Department, John
Wiley & Sons Ltd, The Atrium, Southern Gate, Chichester, West Sussex PO19 8SQ, England, or emailed to
permreq@wiley.co.uk, or faxed to (+44) 1243 770620.

Designations used by companies to distinguish their products are often claimed as trademarks. All brand names
and product names used in this book are trade names, service marks, trademarks or registered trademarks of
their respective owners. The Publisher is not associated with any product or vendor mentioned in this book.

This publication is designed to provide accurate and authoritative information in regard to the subject matter
covered. It is sold on the understanding that the Publisher is not engaged in rendering professional services. If
professional advice or other expert assistance is required, the services of a competent professional should be
sought.

Other Wiley Editorial Offices

John Wiley & Sons Inc., 111 River Street, Hoboken, NJ 07030, USA

Jossey-Bass, 989 Market Street, San Francisco, CA 94103-1741, USA

Wiley-VCH Verlag GmbH, Boschstr. 12, D-69469 Weinheim, Germany

John Wiley & Sons Australia Ltd, 42 McDougall Street, Milton, Queensland 4064, Australia

John Wiley & Sons (Asia) Pte Ltd, 2 Clementi Loop #02-01, Jin Xing Distripark, Singapore 129809

John Wiley & Sons Canada Ltd, 22 Worcester Road, Etobicoke, Ontario, Canada M9W 1L1

Wiley also publishes its books in a variety of electronic formats. Some content that appears
in print may not be available in electronic books.

Library of Congress Cataloging-in-Publication Data

Stocker, Alan.
 Analog VLSI Circuits for the perception of visual motion / Alan Stocker.
    p. cm.
 Includes bibliographical references and index.
 ISBN-13: 978-0-470-85491-4 (cloth : alk. paper)
 ISBN-10: 0-470-85491-X (cloth : alk. paper)
 1. Computer vision. 2. Motion perception (Vision)–Computer simulation.
 3. Neural networks (Computer science) I. Title.
 TA1634.S76 2006
 006.3′7–dc22
                   2005028320

British Library Cataloguing in Publication Data

A catalogue record for this book is available from the British Library

ISBN-13: 978-0-470-85491-4
ISBN-10: 0-470-85491-X

Typeset in 10/12pt Times by Laserwords Private Limited, Chennai, India
Printed and bound in Great Britain by Antony Rowe Ltd, Chippenham, Wiltshire
This book is printed on acid-free paper responsibly manufactured from sustainable forestry
in which at least two trees are planted for each one used for paper production.

What I cannot create, I do not understand.

(Richard P. Feynman – last quote on the blackboard in his office at Caltech when he died in 1988.)

Contents

Foreword

Although we are now able to integrate many millions of transistors on a single chip, our ideas of how to use these transistors have changed very little from the time when John von Neumann first proposed the global memory access, single processor architecture for the programmable serial digital computer. That concept has dominated the last half century, and its success has been propelled by the exponential improvement of hardware fabrication methods reflected in Moore's Law. However, this progress is now reaching a barrier in which the cost and technical problems of constructing CMOS circuits at ever smaller feature sizes is becoming prohibitive. In future, instead of taking gains from transistor count, the hardware industry will explore how to use the existing counts more effectively by the interaction of multiple general and specialist processors. In this way, the computer industry is likely to move toward understanding and implementing more brain-like architectures.

Carver Mead, of Caltech, was one of the pioneers who recognized the inevitability of this trend. In the 1980s he and his collaborators began to explore how integrated hybrid analog–digital CMOS circuits could be used to emulate brain-style processing. It has been a hard journey. Analog computing is difficult because the physics of the material used to construct the machine plays an important role in the solution of the problem. For example, it is difficult to control the physical properties of sub-micron-sized devices such that their analog characteristics are well matched. Another problem is that unlike the bistable digital circuits, analog circuits have no inherent reference against which signal errors can be restored. So, at first sight, it appears that digital machines will always have an advantage over analog ones when high precision and signal reliability are required.

But why are precision and reliability required? It is indeed surprising that the industry insists on developing technologies for precise and reliable computation, despite the fact that brains, which are much more effective than present computers in dealing with real-world tasks, have a data precision of only a few bits and noisy communications.

One factor underlying the success of brains lies in their use of constraint satisfaction. For example, it is likely that the fundamental Gestalt Laws of visual perceptual grouping observed in humans arise from mechanisms that resolve and combine the aspects of an image that cohere from those that do not. These mechanisms rapidly bootstrap globally coherent solutions by quickly satisfying local consistency conditions. Consistency depends on relative computations such as comparison, interpolation, and error feedback, rather than absolute precision. And, this style of computation is suitable for implementation in densely parallel hybrid CMOS circuits.

The relevance of this book is that it describes the theory and practical implementation of constraint satisfaction networks for motion perception. It also presents a principled development of a series of analog VLSI chips that go some way toward the solution of some difficult problems of visual perception, such as the Aperture Problem, and Motion Segmentation.

These classical problems have usually been approached by algorithms, and simulation, suitable for implementation only on powerful digital computers. Alan Stocker's approach has been to find solutions suitable for implementation on a single or very small number of electronic chips that are composed predominantly of analog circuitry, and that process their visual input in real time. His solutions are elegant, and practically useful. The aVLSI design, fabrication, and subsequent analysis have been performed to the highest standards. Stocker discusses each of these phases in some detail, so that the reader is able to gain considerable practical benefit from the author's experience.

Stocker also makes a number of original contributions in this book. The first is his extension of the classical Horn and Schunck algorithm for estimation of two-dimensional optical flow. This algorithm makes use of a brightness and a smoothness constraint. He has extended the algorithm to include a 'bias constraint' that represents the expected motion in case the visual input signal is unreliable or absent. The second is the implementation of this algorithm in a fully functional aVLSI chip. And the third is the implementation of a chip that is able to perform piece-wise smooth optical flow estimation, and so is able (for example) to segment two adjacent pattern fields that have a motion discontinuity at their common boundary. The optical flow field remains smooth within each of the segmented regions.

This book presents a cohesive argument on the use of constraint satisfaction methods for approximate solution of computationally hard problems. The argument begins with a useful and informed analysis of the literature, and ends with the fine example of a hybrid motion-selection chip. This book will be useful to those who have a serious interest in novel styles of computation, and the special purpose hardware that could support them.

Rodney J. Douglas   Zürich, Switzerland

Preface

It was 1986 when John Tanner and Carver Mead published an article describing one of the first analog VLSI visual motion sensors. The chip proposed a novel way of solving a computational problem by a collective parallel effort amongst identical units in a homogeneous network. Each unit contributed to the solution according to its own interests and the final outcome of the system was a collective, overall optimal, solution. When I read the article for the first time ten years later, this concept did not lose any of its appeal. I was immediately intrigued by the novel approach and was fascinated enough to spend the next few years trying to understand and improve this way of computation - despite being told that the original circuit never really worked, and in general, this form of computation was not suited for aVLSI implementations.

Luckily, those people were wrong. Working on this concept of collective computation did not only lead to extensions of the original circuit that actually work robustly under real-world conditions, it also provided me with the intuition and motivation to address fundamental questions in understanding biological neural computation. Constraint satisfaction provides a clear way of solving a computational problem with a complex dynamical network. It provides a motivation for the behavior of such systems by defining the optimal solution and dynamics for a given task. This is of fundamental importance for the understanding of complex systems such as the brain. Addressing the question *what* the system is doing is often not sufficient because of its complexity. Rather, we must also address the functional motivation of the system: *why* is the system doing what it does?

Now, another ten years later, this book summarizes some of my personal development in understanding physical computation in networks, either electronic or neural. This book is intended for physicists, engineers and computational biologists who have a keen interest in the computational question in physical systems. And if this book finally inspires a young graduate student to try to understand complex computational systems and the building of computationally efficient devices then I am very content – even if it takes another ten years for this to happen.

Acknowledgments

I am grateful to many people and institutions that have allowed me to pursue my work with such persistence and great scientific freedom. Foremost I want to thank my former advisor, Rodney Douglas, who provided me with a fantastic scientific environment in which many of the ideas originated that are now captured in this book. I am grateful for his encouragement and support during the writing of the book. Most of the circuits developments were performed when I was with the Institute of Neuroinformatics, Zürich Switzerland. My thanks go to all members of the institute at that time, and in particular to the late Jörg Kramer who introduced me to analog circuits design. I also want to thank the Swiss government, the Körber foundation, and the Howard Hughes Medical Institute for their support during the development and writing of this book.

Many colleagues and collaborators had a direct influence on the final form of this book by either working with me on topics addressed in this book or by providing invaluable suggestions and comments on the manuscript. I am very thankful to know and interact with such excellent and critical minds. These are, in alphabetical order: Vlatko Becanovic, Tobias Delbrück, Rodney Douglas, Ralph Etienne-Cummings, Jakob Heinzle, Patrik Hoyer, Giacomo Indiveri, Jörg Kramer, Nicole Rust, Bertram Shi, and Eero Simoncelli.

Writing a book is a hard optimization problem. There are a large number of constraints that have to be satisfied optimally, many of which are not directly related to work or the book itself. And many of these constraints are contradicting. I am very grateful to my friends and my family who always supported me and helped to solve this optimization problem to the greatest possible satisfaction.

Website to the book

There is a dedicated on-line website accompanying this book where the reader will find supplementary material, such as additional illustrations, video-clips showing the real-time output of the different visual motion sensor and so forth. The address is http://wiley.com/go/analog

The website will also contain updated links to related research projects, conferences and other on-line resources.

1

Introduction

Our world is a visual world. Visual perception is by far the most important sensory process by which we gather and extract information from our environment. Light reflected from objects in our world is a very rich source of information. Its short wavelength and high transmission speed allow us a spatially accurate and fast localization of reflecting surfaces. The spectral variations in wavelength and intensity in the reflected light resemble the physical properties of object surfaces, and provide means to recognize them. The sources that light our world are usually inhomogeneous. The sun, our natural light source, for example, is in good approximation a point source. Inhomogeneous light sources cause shadows and reflectances that are highly correlated with the shape of objects. Thus, knowledge of the spatial position and extent of the light source enables further extraction of information about our environment.

Our world is also a world of motion. We and most other animals are moving creatures. We navigate successfully through a dynamic environment, and we use predominantly visual information to do so. A sense of motion is crucial for the perception of our own motion in relation to other moving and static objects in the environment. We must predict accurately the relative dynamics of objects in the environment in order to plan appropriate actions. Take for example the following situation that illustrates the nature of such a perceptual task: the goal-keeper of a football team is facing a direct free-kick toward his goal.[1] In order to prevent the opposing team from scoring, he needs an accurate estimate of the real motion trajectory of the ball such that he can precisely plan and orchestrate his body movements to catch or deflect the ball appropriately. There is little more than just visual information available to him in order to solve the task. And once he is in motion the situation becomes much more complicated because visual motion information now represents the relative motion between himself and the ball while the important coordinate frame remains

---

[1]There are two remarks to make. First, "football" is referred to as the European-style football, also called "soccer" elsewhere. Second, there is no gender-specific implication here; a male goal-keeper was simply chosen so-as to represent the sheer majority of goal-keepers on earth. In fact, I particularly would like to include non-human, artificial goal-keepers as in robotic football (RoboCup [Kitano et al. 1997]).

Analog VLSI Circuits for the Perception of Visual Motion   A. A. Stocker
© 2006 John Wiley & Sons, Ltd

static (the goal). Yet, despite its difficulty, with appropriate training some of us become astonishingly good at performing this task.

High performance is important because we live in a highly competitive world. The survival of the fittest applies to us as to any other living organism, and although the fields of competition might have slightly shifted and diverted during recent evolutionary history, we had better catch that free-kick if we want to win the game! This competitive pressure not only promotes a visual motion perception system that can determine quickly what is moving where, in which direction, and at what speed; but it also forces this system to be efficient. Efficiency is crucial in biological systems. It encourages solutions that consume the smallest amount of resources of time, substrate, and energy. The requirement for efficiency is advantageous because it drives the system to be quicker, to go further, to last longer, and to have more resources left to solve and perform other tasks at the same time. Our goal-keeper does not have much time to compute the trajectory of the ball. Often only a split second determines a win or a defeat. At the same time he must control his body movements, watch his team-mates, and possibly shout instructions to the defenders. Thus, being the complex sensory-motor system he is, he cannot dedicate all of the resources available to solve a single task.

Compared to human perceptual abilities, nature provides us with even more astonishing examples of efficient visual motion perception. Consider the various flying insects that navigate by visual perception. They weigh only fractions of grams, yet they are able to navigate successfully at high speeds through a complicated environments in which they must resolve visual motions up to 2000 deg/s. [O'Carroll et al. 1996] – and this using only a few drops of nectar a day.

1.1   Artificial Autonomous Systems

What applies to biological systems applies also to a large extent to any artificial autonomous system that behaves freely in a real-world[2] environment. When humankind started to build artificial autonomous systems, it was commonly accepted that such systems would become part of our everyday life by the year 2001. Numberless science-fiction stories and movies have encouraged visions of how such agents should behave and interfere with human society. Although many of these scenarios seem realistic and desirable, they are far from becoming reality in the near future. Briefly, we have a rather good sense of what these agents should be capable of, but we are not able to construct them yet. The (semi-)autonomous rover of NASA's recent Mars missions,[3] or demonstrations of artificial pets,[4] confirm that these fragile and slow state-of-the-art systems are not keeping up with our imagination.

Remarkably, our progress in creating artificial autonomous systems is substantially slower than the general technological advances in recent history. For example, digital microprocessors, our dominant computational technology, have exhibited an incredible development. The integration density literally exploded over the last few decades, and so did

---

[2]The term *real-world* is coined to follow an equivalent logic as the term *real-time*: a real-world environment does not really have to be the "real" world but has to capture its principal characteristics.

[3]*Pathfinder 1997, Mars Exploration Rovers 2004*: http://marsprogram.jpl.nasa.gov

[4]e.g. *AIBO* from SONY: http://www.sony.net/Products/aibo/

the density of computational power [Moore 1965]. By contrast, the vast majority of the predicted scenarios for robots have turned out to be hopelessly unrealistic and over-optimistic.
Why?

In order to answer this question and to understand the limitations of traditional
approaches, we should recall the basic problems faced by an autonomously behaving,
cognitive system. By definition, such a system perceives, takes decisions, and plans actions
on a cognitive level. In doing so, it expresses some degree of intelligence. Our goal-keeper
knows exactly what he has to do in order to defend the free-kick: he has to concentrate on
the ball in order to estimate its trajectory, and then move his body so that he can catch or
deflect the ball. Although his reasoning and perception are cognitive, the immanent interaction between him and his environment is of a different, much more physical kind. Here,
photons are hitting the retina, and muscle-force is being applied to the environment. Fortunately, the goalie is not directly aware of all the individual photons, nor is he in explicit
control of all the individual muscles involved in performing a movement such as catching a
ball. The goal-keeper has a nervous system, and one of its many functions is to instantiate
a *transformation layer* between the environment and his cognitive mind. The brain reduces
and preprocesses the huge amount of noisy sensory data, categorizes and extracts the relevant information, and translates it into a form that is accessible to cognitive reasoning (see
Figure 1.1). This is the process of perception. In the process of action, a similar yet inverse
transformation must take place. The rather global and unspecific cognitive decisions need
to be resolved into a finely orchestrated ensemble of motor commands for the individual
muscles that then interact with the environment. However, the process of action will not
be addressed further in this book.

At an initial step perception requires sensory transduction. A sensory stage measures the
physical properties of the environment and represents these measurements in a signal the

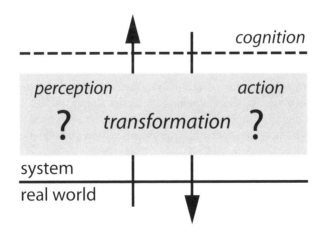

Figure 1.1 *Perception and action.*
Any cognitive autonomous system needs to transform the physical world through perception
into a cognitive syntax and – vice versa – to transform cognitive language into action. The
computational processes and their implementation involved in this transformation are little
understood but are the key factor for the creation of efficient, artificial, autonomous agents.

rest of the system can process. It is, however, clear that sensory transduction is not the only transformation process of perception. Because if it were, the cognitive abilities would be completely overwhelmed with detailed information. As pointed out, an important purpose of perception is to reduce the raw sensory data and extract only the relevant information. This includes tasks such as object recognition, coordinate transformation, motion estimation, and so forth. Perception is the *interpretation* of sensory information with respect to the perceptual goal. The sensory stage is typically limited, and sensory information may be ambiguous and is usually corrupted by noise. Perception, however, must be robust to noise and resolve ambiguities when they occur. Sometimes, this includes the necessity to fill in missing information according to expectations, which can sometimes lead to wrong interpretations: most of us have experienced certainly one or more of the many examples of perceptual illusions.

Although not described in more detail at this point, perceptional processes often represent large computational problems that need to be solved in a small amount of time. It is clear that the efficient implementation of solutions to these tasks crucially determines the performance of the whole autonomous system. Traditional solutions to these computational problems almost exclusively rely on the digital computational architecture as outlined by von Neumann [1945].[5] Although solutions to all computable problems can be implemented in the von Neumann framework [Turing 1950], it is questionable that these implementations are equally efficient. For example, consider the simple operation of adding two analog variables: a digital implementation of addition requires the digitization of the two values, the subsequent storage of the two binary strings, and a register that finally performs the binary addition. Depending on the resolution, the electronic implementation can use up to several hundred transistors and require multiple processing cycles [Reyneri 2003]. In contrast, assuming that the two variables are represented by two electrical currents flowing in two wires, the same addition can be performed by simply connecting the two wires and relying on Kirchhoff's current law.

The von Neumann framework also favors a particular philosophy of computation. Due to its completely discrete nature, it forces solutions to be dissected into a large number of very small and sequential processing steps. While the framework is very successful in implementing clearly structured, exact mathematical problems, it is unclear if it is well suited to implement solutions for perceptual problems in autonomous systems. The computational framework and the computational problems simply do not seem to match: on the one hand the digital, sequential machinery only accepts defined states, and on the other hand the often ambiguous, perceptual problems require parallel processing of continuous measures.

It may be that digital, sequential computation is a valid concept for building autonomous artificial systems that are as powerful and intelligent as we imagine. It may be that we can make up for its inefficiency with the still rapidly growing advances in digital processor technology. However, I doubt it. But how amazing would the possibilities be if we could find and develop a more efficient implementation framework? There must be a different, more efficient way of solving such problems – and that's what this book is about. It aims to demonstrate another way of thinking of solutions to these problems and implementing

---

[5]In retrospect, it is remarkable that from the very beginning, John von Neumann referred to his idea of a computational device as an explanation and even a model of how biological neural networks process information.

them. And, in fact, the burden to prove that there are indeed other and much more efficient ways of computation has been carried by someone else – nature.

1.2   Neural Computation and Analog Integrated Circuits

Biological neural networks are examples of wonderfully engineered and efficient computational systems. When researchers first began to develop mathematical models for how nervous systems actually compute and process information, they very soon realized that one of the main reasons for the impressive computational power and efficiency of neural networks is the collective computation that takes place among their highly connected neurons. In one of the most influential and ground-breaking papers, which arguably initiated the field of computational neuroscience, McCulloch and Pitts [1943] proved that any finite logical expression can be realized by networks of very simple, binary computational units. This was, and still is, an impressive result because it demonstrated that computationally very limited processing units can perform very complex computations when connected together. Unfortunately, many researchers concluded therefore that the brain is nothing more than a big logical device – a digital computer. This is of course not the case because McCulloch and Pitts' model is not a good approximation of our brain, which they were well aware of at the time their work was published.

Another key feature of neuronal structures – which was neglected in McCulloch and Pitts' model – is that they make computational use of their intrinsic physical properties. Neural computation is physical computation. Neural systems do not have a centralized structure in which memory and hardware, algorithm and computational machinery, are physically separated. In neurons, the function is the architecture – and vice versa. While the bare-bone simple McCulloch and Pitts model approximates neurons to be binary and without any dynamics, real neurons follow the continuous dynamics of their physical properties and underlying chemical processes and are analog in many respects. Real neurons have a cell membrane with a capacitance that acts as a low-pass filter to the incoming signal through its dendrites, they have dendritic trees that non-linearly add signals from other neurons, and so forth. John Hopfield showed in his classical papers [Hopfield 1982, Hopfield 1984] that the dynamics of the model neurons in his networks are a crucial prerequisite to compute near-optimal solutions for hard optimization problems with recurrent neural networks [Hopfield and Tank 1985]. More importantly, these networks are very efficient, establishing the solution within a few characteristic time constants of an individual neuron. And they typically scale very favorably. Network structure and analog processing seem to be two key properties of nervous systems providing them with efficiency and computational power, but nonetheless two properties that digital computers typically do not share or exploit. Presumably, nervous systems are very well optimized to solve the kinds of computational problems that they have to solve to guarantee survival of their whole organism. So it seems very promising to reveal these optimal computational strategies, develop a methodology, and transfer it to technology in order to create efficient solutions for particular classes of computational problems.

It was Carver Mead who, inspired by the course "The Physics of Computation" he jointly taught with John Hopfield and Richard Feynman at Caltech in 1982, first proposed the idea of embodying neural computation in silicon *analog very large-scale integrated (aVLSI)* circuits, a technology which he initially advanced for the development of integrated digital circuits.[6] Mead's book *Analog VLSI and Neural Systems* [Mead 1989] was a sparkling source of inspiration for this new emerging field, often called *neuromorphic* [Mead 1990] or *neuro-inspired* [Vittoz 1989] circuit design. And nothing illustrates better the motivation for the new field than Carver Mead writing in his book: "Our struggles with digital computers have taught us much about how neural computation *is not* done; unfortunately, they have taught us relatively little about how it *is* done."

In the meantime, many of these systems have been developed, particularly for perceptual tasks, of which the *silicon retina* [Mahowald and Mead 1991] was certainly one of the most popular examples. The field is still young. Inevitable technological problems have led now to a more realistic assessment of how quickly the development will continue than in the euphoric excitement of its beginning. But the potential of these neuromorphic systems is obvious and the growing scientific interest is documented by an ever-increasing number of dedicated conferences and publications. The importance of these neuromorphic circuits in the development of autonomous artificial systems cannot be over-estimated.

This book is a contribution to further promote this approach. Nevertheless, it is as much about network computation as about hardware implementation. In that sense it is perhaps closer to the original ideas of Hopfield and Mead than current research. The perception of visual motion thereby only serves as the example task to address the fundamental problems in artificial perception, and to illustrate efficient solutions by means of analog VLSI network implementations. In many senses, the proposed solutions use the same computational approach and strategy as we believe neural systems do to solve perceptual problems. However, the presented networks are not designed to reflect the biological reference as thoroughly as possible. The book carefully avoids using the term *neuron* in any other than its biological meaning. Despite many similarities, silicon aVLSI circuits are bound to their own physical constraints that in many ways diverge from the constraints nervous systems are facing. It does not seem sensible to copy biological circuits as exactly as possible. Rather, this book aims to show how to use basic computational principles that we believe make nervous systems so efficient and apply them to the new substrate and the task to solve.

---

[6]There were earlier attempts to build analog electrical models of neural systems. Fukushima et al. [1970] built an electronic retina from discrete(!) electronic parts. However, only when integrated technology became available were such circuits of practical interest.

2

Visual Motion Perception

Visual motion perception is the process an observer performs in order to extract relative motion between itself and its environment using visual information only. Typically, the observer possesses one or more imaging devices, such as eyes or cameras. These devices sense images that are the two-dimensional projection of the intensity distribution radiating from the surfaces of the environment. When the observer moves relative to its environment, its motion is reflected in the images accordingly. Because of this causal relationship, being able to perceive *image motion* provides the observer with useful information about the relative physical motion. The problem is that the physical motion is only implicitly represented in the spatio-temporal brightness changes reported by the imaging devices. It is the task of visual motion perception to interpret the spatio-temporal brightness pattern and extract image motion in a meaningful way.

This chapter will outline the computational problems involved in the perception of visual motion, and provide a rough concept of how a system for visual motion perception should be constructed. The concept follows an ecological approach. Visual motion perception is considered to be performed by a completely autonomous observer behaving in a real-world environment. Consequently, I will discuss the perceptual process with respect to the needs and requirements of the observer. Every now and then, I will refer also to biological visual motion systems, mostly of primates and insects, because these are examples that operate successfully under real-world conditions.

2.1 Image Brightness

Visual motion perception begins with the acquisition of visual information. The imaging devices of the observer, referred to in the following as *imagers*, allow this acquisition by (i) mapping the visual scene through suitable optics onto a two-dimensional image plane, and (ii) transducing and decoding the projected intensity into appropriate signals that the subsequent (motion) systems can process.

Figure 2.1 schematically illustrates the imaging. The scene consists of objects that are either direct (sun) or indirect (tree) sources of light, and their strength is characterized by the

Analog VLSI Circuits for the Perception of Visual Motion   A. A. Stocker
© 2006 John Wiley & Sons, Ltd

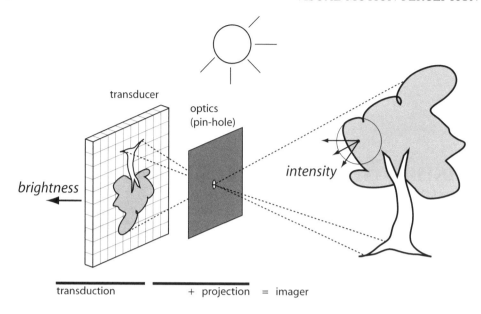

Figure 2.1 *Intensity and brightness.*
Intensity is a physical property of the object while brightness refers to the imager's subjective measure of the projected intensity. Brightness accounts for the characteristics of the projection and transduction process. Each pixel of the imager reports a brightness value at any given time. The ensemble of pixel values represents the image.

total power of their radiation, called radiant flux and measured in watts [W]. If interested in the perceptual power, the flux is normalized by the spectral sensitivity curves of the human eye. In this case, it is referred to as luminous flux and is measured in lumen [lm]. For example, a radiant flux of 1 W at a wavelength of 550 nm is approximately 680 lm, whereas at 650 nm it is only 73 lm. The radiation emitted by these objects varies as a function of direction. In the direction of the imager, each point on the objects has a particular *intensity*, defined as the flux density (flux per solid angle steradian [W/sr]). It is called luminous intensity if converted to perceptual units, measured in candelas [1 cd = 1 lm/sr]. In the current context, however, the distinction between radian and luminous units is not important. After all, a spectral normalization only make sense if it was according to the spectral sensitivity of the particular imager. What is important to note is that intensity is an *object property*, thus is independent of the characteristics of the imager processing it.

The optics of the imager in Figure 2.1, in this case a simple pin-hole, create a projection of the intensity distribution of the tree onto a transducer. This transducer, be it a CCD chip, a biological or artificial retina, consists of an array of individual picture elements, in short *pixels*. The intensity over the size of each pixel is equal to the radiance [W/sr/m^2] (resp. luminance) of the projected object area. Because radiance is independent of the distance, knowing the pixel size and the optical pathway alone would, in principle, be sufficient to extract the intensity of the object.

Figure 2.2 *Brightness is a subjective measure.*
The two small gray squares appear to differ in brightness although, assuming a homogeneous illumination of this page, the intensities of each square are identical. The perceptual difference emerges because the human visual system is modulated by spatial context, where the black background makes the gray square on the right appear brighter than the one on the left. The effect is strongest when observed at about arm-length distance.

Brightness, on the other hand, has no SI units. It is a *subjective* measure to describe how bright an object appears. Brightness reflects the radiance of the observed objects but is strongly affected by contextual factors such as, for example, the background of the visual scene. Many optical illusions such as the one in Figure 2.2 demonstrate that these factors are strong; humans have a very subjective perception of brightness.

An imager is only the initial processing stage of visual perception and hardly operates on the notion of objects and context. It simply transforms the visual environment into an image, which represents the spatially sampled measure of the radiance distribution in the visual scene observed. Nevertheless, it is sensible to refer to the image as representing a brightness distribution of the visual scene, to denote the dependence of the image transduction on the characteristics of the imager. The image no longer represents the radiance distribution that falls onto the transducer. In fact, a faithful measurement of radiance is often not desirable given the huge dynamic range of visual scenes. An efficient imager applies local preprocessing such as compression and adaptation to save bandwidth and discard visual information that is not necessary for the desired subsequent processing. Imaging is the first processing step in visual motion perception. It can have a substantial influence on the visual motion estimation problem.

Throughout this book, an image is always referred to represent the output of an imager which is a subjective measure of the projected object intensity. While intensity is a purely object related physical property, and radiance is what the imager measures, brightness is what it reports.

2.2 Correspondence Problem

Image motion, and the two-dimensional projection of the three-dimensional motion (in the following called *motion field*), are generally not identical [Marr 1982, Horn 1986, Verri and Poggio 1989]. The reason for this difference is that individual locations on object surfaces in space are characterized only by their light intensities. Thus, the brightness distribution in the image can serve only as an indirect measure of the motion field. And unfortunately, of course, brightness is not a unique label for individual points in space. This is the essential problem of visual motion perception, given that its goal is to extract the motion field as faithfully as possible. In fact, this is the essential problem of any visual processing. The problem is usually called the *correspondence problem* [Ullman 1979] and has become prominent with the problem of depth estimation using stereopsis [Marr and Poggio 1976].

Figure 2.3 illustrates the correspondence problem, together with some of its typical instantiations. One common example is a rotating but translationally static sphere that has no texture. Its physical motion does not induce any image motion. Another example is transparent motion, produced by translucent overlapping stimuli moving past one another in different directions. Such transparent motion can produce an ambiguous spatio-temporal brightness pattern in the image, such as image motion that does not coincide with any of the motion fields of the two individual stimuli. In Figure 2.3b, image motion appears to be downward, whereas the physical motion of the individual stimuli is either horizontally to the right or to the left. Note that the resulting image motion is not the average motion of the two stimuli, which one might naively presume. Another instantiation of the correspondence problem is the so-called *aperture problem* [Marr 1982, Hildreth 1983] illustrated in Figure 2.3c. It describes the ambiguity in image motion that necessarily occurs (in the limit) when observing the image through a small aperture. I will discuss the aperture problem in more detail when addressing motion integration. The spatio-temporal brightness changes in the image do not have to originate from direct physical motion within the environment. Consider for example a static scene illuminated by a light beam from outside the visual field of the observer such that objects in the field cast shadows. If the light beam moves, the shadows will also move and induce image motion, although there is no physical motion within the visual field. In general, reflectance properties and scene illumination are dominant factors that determine how well image motion matches the motion field.

What becomes clear from the above examples is that the perception of image motion is an estimation rather than a direct measurement process. The correspondence problem requires that the estimation process uses additional information in order to resolve the inherent ambiguities, and to provide a clean and unambiguous percept. The percept is an interpretation of the visual information (the spatio-temporal brightness distribution) using prior assumptions about the source of the image motion. The estimation process is also determined by the needs and quality requirements of its estimates. Assume that image motion estimation is a subtask of a complex system (such as the goal-keeper in the previous chapter). Then the functionality and purpose of the whole system influence the way the visual motion information must be interpreted. Perhaps, to serve its task, the complete system does not require a very accurate estimation of image motion in terms of a faithful

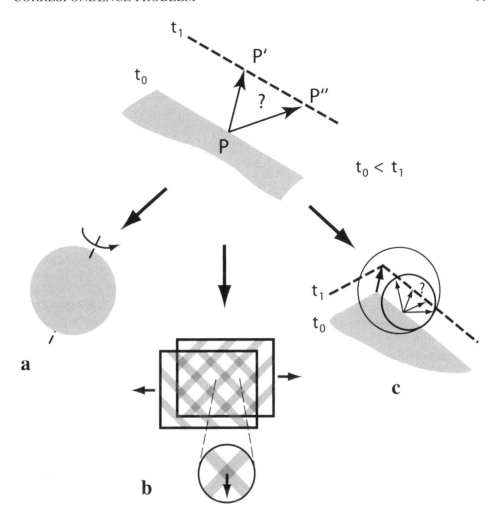

Figure 2.3 *Correspondence problem.*
Intensity is an ambiguous label for individual locations on an object in space. Consequently, image brightness is ambiguous as well. For example, the point P does not have an individual brightness signature among other points on the moving edge. Thus, a displacement of the brightness edge (illustrated by the dashed line) does not allow the observer to determine if P' or P'' is the new location of P. This correspondence problem expresses itself prominently as unperceived motion of visually unstructured objects (a), as transparent motion (b), or as the aperture problem (c).

estimate of the motion field. In this case it would be inefficient to estimate information that is not required. Some interesting and challenging thoughts on this matter can be found in Gibson [1986]. In any case, a system that estimates image motion is performing a very subjective task, constrained by the assumptions and models it applies.

This raises the question of how to define a correct benchmark for visual motion systems. Is a best possible estimation of the motion field the preferred goal? Well, as said before, this depends on the needs of the entire system. In general, it seems sensible to extract the motion field as well as possible [Little and Verri 1989, Barron et al. 1994], although we have to keep in mind that non-physical motion such as the moving shadows induces image motion that can also provide useful information to the observer, namely that the light source is moving. However, extracting the exact motion field requires the resolution of all potential ambiguities and this implies that the system has an "understanding" of its environment. Basically, it requires a complete knowledge of the shapes and illumination of the environment, or at least all necessary information to extract those. The problem is then, how does the system acquire this knowledge of its environment if not through perception of which visual motion perception is a part? This is a circular argument. Clearly, if the system must know everything about its environment in order to perform perception, then perception seems to be redundant.

Consequently, visual motion perception must not make too strong assumptions about the observed physical environment. Strong assumptions can result in discarding or misinterpreting information that later processing stages might need, or would be able to interpret correctly. Therefore a visual motion perception system must mainly rely on the *bottom-up* visual information it receives. Yet, at the same time, it should allow us to incorporate *top-down* input from potential higher level stages to help refine its estimate of image motion, which then in return provides better information for such later stages. We see that such recurrent processing is a direct reflection of the above circular argument.

2.3   Optical Flow

Before I discuss computational methods to extract image motion, I should address briefly the representation of such image motion. In principle, the representation or encoding can be arbitrary. But it is better to adapt the representation to the computational machinery, the affordable communication bandwidth of the system, and the needs of later processing stages. A dense, and so very flexible form of representation is a vector field $\vec{v}(x, y, t) = (u(x, y, t), v(x, y, t))$ that characterizes the direction and speed of image motion at each particular time and image location. To improve readability, space and time dependence of u and v will not be explicitly written out in subsequent notations. Such a vector field is referred to as optic or *optical flow* [Gibson 1950]. Unfortunately, the meaning of optical flow differs somewhat, depending on the scientific field in which it is used. Perceptual psychologists (such as James Gibson or Jan Koenderink [Koenderink 1986], to name two) typically define optical flow as the kind of image motion that originates from a relative motion between an observer and a static but cluttered environment. Thus the resulting image motion pattern provides mainly information about global motion of the observer, e.g. ego-motion. In computer vision and robotics, however, the definition of optical flow is much more generic. It is defined as the representation of the local image motion originating from any relative motion between an observer and its potentially non-static environment [Horn and Schunck 1981, Horn 1986]. I will refer to optical flow in this latter, more generic sense. Any concurrent interpretation of the pattern of optical flow as induced by specific relative motions is assumed to take place in a later processing stage.

Nature seems to concur with this sense. We know that particular areas of a primate's visual cortex encode local image motion in a manner similar to optical flow. For example, the dominant motion-sensitive area *medial temporal* (MT) in macaque monkeys is retino-topically organized in columns of directional and speed-sensitive neurons in a way that is very similar to the orientation columns in the primary visual cortex V1 [Movshon et al. 1985] (see [Lappe 2000] for a review).

2.4 Matching Models

A basic assumption to extract visual motion is that visual information is preserved over time, but possibly displaced in space. That is, it is assumed that a feature in the image at some point in time will also be present somewhere in the image at a later time. Given this assumption, a matching procedure that measures the image displacements of features over time could provide a quantitative measure for image motion. This assumption is simplistic and often violated, for example when occlusions occur (see Figure 2.3b) or for any kind of perspective motion that is not fronto-parallel to the imager (see e.g. the radial flow example in Figure 2.4b). On a small scale (time and space), however, such perspective motion can be well approximated as translational fronto-parallel motion. This is also true for rotational or non-rigid motion. Consequently, I will consider matching only under translational displacements. In the following two classes of methods are distinguished.

2.4.1 Explicit matching

The first class includes all those methods that apply matching explicitly: they first extract features and then track them in time. The displacement per time unit in image space serves as the measure for image motion. Features can be thought of as being extracted on different levels of abstraction, from the raw brightness patches to low-level features like edges up to complete objects. Typically, these methods can be described by the following matching operation:

$$M(x, y; \Delta x, \Delta y) = \phi\big(E(x, y, t), E(x + \Delta x, y + \Delta y, t + \Delta t)\big) \qquad (2.1)$$

where E is the representation of some features in image space and time, Δx and Δy are the translational displacements in image space, and ϕ is some correlation function that is maximal if $E(x, y, t)$ and $E(x + \Delta x, y + \Delta y, t + \Delta t)$ are most similar [Anandan 1989, Bülthoff et al. 1989]. The task reduces to finding Δx and Δy such that $M(x, y; \Delta x, \Delta y)$ is maximal, hence

$$(\Delta x, \Delta y) := \arg \max_{\Delta x, \Delta y} M(x, y; \Delta x, \Delta y). \qquad (2.2)$$

Given the relative time difference Δt, the image velocity is proportional to the displacements. Or given the distance, the velocity is inversely proportional to the time difference [Camus 1994].

Unfortunately, maximizing M can be an ambiguous problem and so mathematically ill-posed.[1] That is, there may be several solutions maximizing M. We are faced with the

---

[1]See Appendix A

correspondence problem. Applying a competitive selection mechanism may always guarantee a solution. However, in this case the correspondence problem is just hidden in such a way that the selection of the motion estimate in an ambiguous situation is either a random choice amongst the several maxima, or driven by noise and therefore ill-conditioned. Neither of these two decision mechanisms seem desirable. The extraction and tracking of higher level spatial features does not circumvent the problem. In this case, the correspondence problem is just partially shifted to the extraction process, which is itself ill-posed [Bertero et al. 1987], and partially to the tracking depending on the level of feature extraction; for example, tracking edges might be ambiguous if there is occlusion.

There are two other reasons against having a complex spatial feature extraction stage preceding image motion estimation. First, fault tolerance is decreased, because once a complex feature is misclassified the motion information of the entire feature is wrong. In contrast, outliers on low-level features can be discarded, usually by some confidence measure, or simply averaged out. Second, because spatial feature extraction occurs before any motion is computed, motion cannot serve as a cue to enhance the feature extraction process, which seems to be an inefficient strategy.

2.4.2   Implicit matching

The second class contains those methods that rely on the continuous interdependence between image motion and the spatio-temporal pattern observed at some image location. The matching process is implicit.

Gradient-based methods also assume that the brightness of an image point remains constant while undergoing visual motion [Fennema and Thompson 1979]. Let $E(x, y, t)$ describe the brightness distribution in the image on a Cartesian coordinate system. The Taylor expansion of $E(x, y, t)$ leads to

$$E(x, y, t) = E(x_0, y_0, t_0) + \frac{\partial E}{\partial x}\mathrm{d}x + \frac{\partial E}{\partial y}\mathrm{d}y + \frac{\partial E}{\partial t}\mathrm{d}t + \epsilon \qquad (2.3)$$

where ϵ contains higher order terms that are neglected. Assuming that the brightness of a moving image point is constant, that is $E(x, y, t) = E(x_0, y_0, t_0)$, and dividing by $\mathrm{d}t$ leads to

$$\frac{\partial}{\partial x}E(x, y, t)\, u + \frac{\partial}{\partial y}E(x, y, t)\, v + \frac{\partial}{\partial t}E(x, y, t) = 0 \qquad (2.4)$$

where $u = \mathrm{d}x/\mathrm{d}t$ and $v = \mathrm{d}y/\mathrm{d}t$ are the components of the local optical flow vector $\vec{v} = (u, v)$.

Equation (2.4) is called the *brightness constancy equation*,[2] first introduced by Fennema and Thompson [1979] (see also [Horn and Schunck 1981]). Obviously, the brightness constancy equation is almost never exactly true. For example, it requires that every change in brightness is due to motion; that object surfaces are opaque and scatter light equally in all directions; and that no occlusions occur. Many of these objections are inherent problems of the estimation of optical flow. Even if brightness constancy were to hold perfectly, we could not extract a dense optical flow field because the single Equation (2.4) contains two unknowns, u and v. Consequently, the computational problem of estimating local visual

---

[2]Sometimes it is also referred to as the *motion constraint equation*.

motion using the brightness constancy alone is ill-posed, as are many other tasks in visual processing [Poggio et al. 1985]. Nevertheless, the brightness constancy equation grasps the basic relation between image motion and brightness variations, and has proven to be a valid first-order model. Equation (2.4) is also a formal description of the aperture problem.

There is no fundamental reason that restricts the assumption of constancy to brightness. Visual information can be preprocessed by a local stationary operator. Then, an equivalent *general image constancy equation* relates visual information to visual motion where $E(x, y, t)$ in (2.4) is being replaced by the output of the preprocessing operator. Requirements are that the output of the preprocessor is differentiable, and thus that the spatio-temporal gradients are defined. For example, Fleet and Jepson [1990] applied a spatial Gabor filter to the image brightness and then assumed constancy in the phase of the filter output. Phase is considered to be much less affected by variations of scene illumination than image brightness and assuming constancy in phase seems to provide more robust optical flow estimates. In Chapter 5, I will readdress this issue of a general image constancy constraint in the context of the characteristics of the phototransduction circuits of aVLSI implementations.

A second group of implicit matching models characterize the spatio-temporal nature of visual motion by the response to spatially and temporally oriented filters. A relatively simple model in one spatial dimension was proposed by Hassenstein and Reichardt [1956] based on their studies of the visual motion system of the beetle species *Chlorophanus*. This *correlation method*, which turns out to be a common mechanism in insect motion vision, correlates the temporally low-pass-filtered response of a spatial feature detector with the temporally high-pass-filtered output from its neighbor. The correlation will be maximal if the observed stimulus matches the time constant of the low-pass filter. A similar arrangement was found also in the rabbit retina [Barlow and Levick 1965]. In this case, the output of a feature detector is inhibited by the delayed output of its neighbor located in the preferred moving direction. A stimulus moving in the preferred direction will elicit a response that is suppressed by the delayed output of the neighboring detector (in moving direction). In the null direction, the detector is inhibited by its neighbor if the stimulus matches the time constant of the delay element. Correlation methods do not explicitly report velocity. Their motion response is phase-dependent. It has been shown that correlation models are computationally equivalent to the first stage of a more general family of models [Van Santen and Sperling 1984]. These *motion energy models* [Watson and Ahumada 1985, Adelson and Bergen 1985] apply odd-and even-type Gabor filters in the spatio-temporal frequency domain so that their combined output is approximately phase-invariant, and reaches a maximum for stimuli of a particular spatial and temporal frequency. Many of these filters tuned to different combinations of spatial and temporal frequencies are integrated to provide support for a particular image velocity [Heeger 1987a, Grzywacz and Yuille 1990, Simoncelli and Heeger 1998]. Motion energy is usually computed over an extended spatio-temporal frequency range. Therefore, motion energy models are possibly more robust than gradient-based models in a natural visual environment that typically exhibits a broadband spatio-temporal frequency spectrum.

None of the above matching models can entirely circumvent the inherent correspondence problem. The purpose of a matching model is to establish a function of possible optical flow estimates given the visual data. From all imaginable solutions, it constrains a subset of optical flow estimates which is in agreement with the visual data. The matching model does

not have to provide a single estimate of optical flow. On the contrary, it is important that
the matching model makes as few assumptions as necessary, and does not possibly discard
the desired "right" optical flow estimate up-front. The final estimate is then provided by
further assumptions that select one solution from the proposed subset. These additional
assumptions are discussed next.

2.5   Flow Models

Flow models are typically parametric models and reflect the expectations of the observed
type of image motion. These models represent *a priori* assumptions about the environment
that might be formed by adaptation processes on various time-scales [Mead 1990, Rao and
Ballard 1996, Rao and Ballard 1999]. They can include constraints on the type of motion,
the image region in which the model applies, as well as the expected statistics of the image
motion over time.

The more detailed and specified a model is and thus the more accurate it can describe a
particular flow field, the more it lacks generality. The choice of the model is determined by
the image motion expected, the type of motion information required, and the complexity
of the system allowed. Furthermore, depending on their complexity and thus the required
accuracy of the model, flow models can permit a very compact and sparse representation
of image motion. Sparse representations become important, for example, in efficient video
compression standards.

2.5.1   Global motion

First, consider the modeling of the flow field induced by relative movements between
the observer and its environment. If the environment remains stationary, image motion is
directly related to the ego-motion of the observer. In this case, the observer does not have
to perceive its environment as a collection of single objects but rather sees it as a spatio-
temporally structured background that permits it to sense its own motion [Sundareswaran
1991].

As illustrated in Figure 2.4, three fundamental optical flow fields can be associated with
global motion relative to a fronto-parallel oriented background:

- The simplest flow field imaginable results from pure *translational* motion. The
  induced flow field $\vec{v}(x, y, t)$ does not contain any source or rotation. Thus the diver-
  gence $\text{div}(\vec{v})$, the rotation $\text{rot}(\vec{v})$, and the gradient $\text{grad}(\vec{v})$ of the flow field are zero.
  Such global translational motion can be represented by a single flow vector.

- Approaching or receding motion will result in a *radial* flow field that contains a
  single source or sink respectively. An appropriate model requires the divergence to be
  constant. For approaching motion, the origin of the source is called *focus of expansion*
  and signals the heading direction of the observer. Similarly for receding motion, the
  origin of the sink is called *focus of contraction*. A single parameter $c_0 = \text{div}(\vec{v})$ is
  sufficient to describe the flow field where its sign indicates approaching or receding
  motion and its value is a measure for speed.

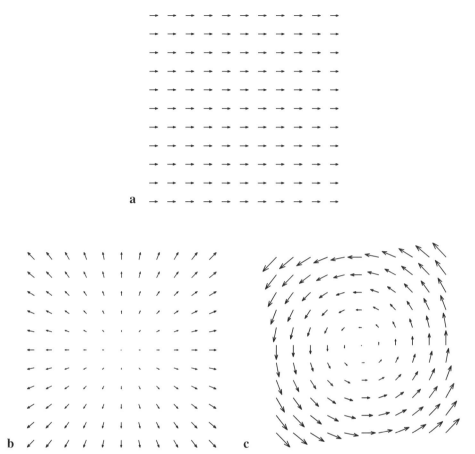

Figure 2.4 *Basic global flow patterns.*
A purely translational (a), radial (b), and rotational (c) flow field.

- Pure rotational motion between the observer and the background at constant distance will induce a *rotational* flow field as shown in Figure 2.4c. The flow field has again no source, but now rotation is present. Clockwise or counter-clockwise rotation can be described by the sign of vector $\vec{c} = \text{rot}(\vec{v})$ pointing either perpendicularly in or out of the image plane. Since $\vec{c}$ is constant in the entire image space, a single parameter is sufficient to describe the flow field.

Many more complicated flow models of affine, projective, or more general polynomial type have been proposed to account for the motion of tilted and non-rigid objects (for a review see [Stiller and Konrad 1999]). Furthermore, there are models that also consider the temporal changes of the flow field such as acceleration [Chahine and Konrad 1995]. However, it is clear that an accurate description of the complete flow field with a single global model containing a few parameters is only possible for a few special cases. Typically, it will oversimplify the visual motion. Nevertheless, for many applications a coarse approximation is sufficient and a global flow model provides a very sparse representation of visual motion. Global motion is generally a robust estimate because it combines visual information over the whole image space.

There is evidence that the visual system in primates also extracts global flow patterns for ego-motion perception [Bremmer and Lappe 1999]. Recent electro-physiological studies show that in regions beyond area MT of the macaque monkey neurons respond to global flow patterns [Duffy and Wurtz 1991a, Duffy and Wurtz 1991b]. In the medial superior Temporal (MST) area neurons are selective to particular global flow patterns such as radial and rotational flow and combinations of it [Lappe et al. 1996]. Furthermore a large fraction of neurons in area MST receive input from MT [Duffy 2000] and vestibular information. These observations suggest that MST is a processing area that occurs after MT in the processing stream and is particularly involved in the perception of ego-motion. Similarly, the fly's visual system has neurons that respond to global flow patterns associated with self-motions. These so-called HS and VS neurons selectively integrate responses from a subset of local elementary motion detectors to respond only to particular global motion patterns [Krapp 2000].

2.5.2 Local motion

Global flow models are sufficient to account for ego-motion and permit the extraction of important variables such as heading direction. Global flow models are well suited to account for visual motion of only one motion source (e.g. ego-motion). However, they usually fail when multiple independent motion sources are present. For this case, more local motion models are required.

A strictly local model[3] is possible, but cannot do a very good job because of the aperture problem. The implications of the aperture problem are illustrated in Figure 2.5a. The figure shows the image of a horizontally moving triangle at two different points in time. Each of the three circles represents an aperture through which local image motion is observed. In the present configuration, each aperture permits only the observation of a local feature of the moving object which in this case is either a brightness edge of a particular spatial orientation (apertures C and B) or the non-textured body of the object (aperture A). The image motion within each aperture is ambiguous and could be elicited by an infinite number of possible local displacements as indicated by the set of vectors. Necessarily, a strictly local model is limited. One possible local model is to choose the shortest from the subset of all possible motion vectors within each aperture (bold arrows). Since the shortest flow vector points perpendicular to the edge orientation, this model is also called the *normal flow model*. Using the gradient-based matching model, the end points of all possible local flow vectors lie on the constraint lines given by the brightness constancy Equation (2.4). The normal flow vector $\vec{v}_n = (u_n, v_n)$ is defined as the point on the constraint line that is closest to the origin (illustrated in Figure 2.5b), thus

$$u = -\frac{E_t E_x}{E_x^2 + E_y^2} \quad \text{and} \quad v = -\frac{E_t E_y}{E_x^2 + E_y^2}. \qquad (2.5)$$

---

[3]In practice, images are spatially sampled brightness distributions. I use *strictly local* to refer to the smallest possible spatial scale, which is the size of a single pixel.

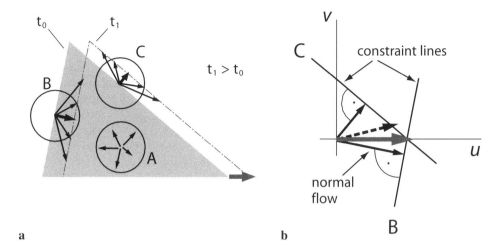

Figure 2.5 *The aperture problem.*
(a) Translational image motion provides locally ambiguous visual motion information. Apertures containing zero-order (aperture A) or first-order (apertures B and C) spatio-temporal brightness patterns do not allow an unambiguous estimate of local motion. (b) Vector averaging of the normal flow field does not lead to the correct global motion (dashed arrow). Instead, only the *intersection-of-constraints (IOC)* provides the correct image motion of the object (bold arrow).

Normal flow can be estimated in apertures B and C, yet Equation (2.5) is not defined for aperture A because the spatio-temporal brightness gradients are zero. An additional constraint is needed to resolve the ambiguity in aperture A. Typically, a threshold is applied to the spatio-temporal brightness gradients below which image motion is assumed to be zero. Another possibility is to include a small constant in the denominator of (2.5). This will be discussed in more detail later in Chapter 4. Under the assumption that all possible local motions occur with equal probability, normal flow is an optimal local motion estimate with respect to the accumulated least-squares error in direction because it represents the mean direction. On the other hand, if the combined error for direction and speed is measured as the dot-product, the accumulated error remains constant, thus each of the possible flow vectors is an equally (sub-)optimal estimate.

Normal flow is a simple model but can still provide an approximate estimate of local image motion. It can be sufficient for solving some perceptual tasks [Huang and Aloimonos 1991]. However, for many tasks it is not. The image motion of objects, for example, typically does not coincide with normal flow. The vector average of the normal flow field along the object contour does not provide the image motion of the object also. This is illustrated in Figure 2.5b: the vector average estimate (dashed arrow) differs significantly from the image motion of the object determined as the unique motion vector that is present in both subsets of possible flow vectors of apertures B and C. This *intersection-of-constraints (IOC)* solution is, of course, no longer strictly local.

Spatial integration

To avoid the complexity of a global model and to overcome the limitations of a strictly local flow model, one can use intermediate models that account for some extended areas in image space. A first possibility is to partition the image space into sub-images of fixed size and then apply a global model for each sub-image. Probably, the image partitions contain enough brightness structure to overcome the aperture problem. According to the model complexity and the partition sizes, image motion can be captured by a relatively small number of parameters. There is obviously a trade-off between partition size and spatial resolution, thus quality, of the estimated image motion. Current video compression standards (MPEG-2 [1997]) typically rely on partitioning techniques with fixed block sizes of 16×16 pixels, assuming an independent translational flow model for each block. Figure 2.6b illustrates such a block-partition approach for a scene containing two moving objects. For the sake of simplicity they are assumed to undergo purely translational motion. A fixed partition scheme will fail when several flow sources are simultaneously present in one partition, which is not the case in this example. It also leads to a discontinuous optical flow field along the partition boundaries. The resulting optical flow estimate is non-isotropic and strongly affected by the predefined regions of support. The advantage of block-partitioning is its sparse representation of motion, and its relatively low computational load, which allow an efficient compression of video streams.

Overlapping kernels [Lucas and Kanade 1981] or isotropic diffusion of visual motion information [Horn and Schunck 1981, Hildreth 1983] can circumvent some of the disadvantages of block-partition approaches. These methods assume that the optical flow varies smoothly over image space, and the model yields smooth optical flow fields as illustrated in Figure 2.6c. Smoothness, however, is clearly violated at object boundaries, where motion discontinuities usually occur.

Neither fixed partitions nor isotropic kernels adequately reflect the extents of arbitrary individual motion sources. Ideally, one would like to have a flow model that selects regions of support that coincide with the outline of the individual motion sources, and then estimates the image motion independently within each region of support. This approach is illustrated in Figure 2.6d. As before, the background is stationary and the two objects are moving independently. The induced flow field of each object is preferably represented by a *separate* translational flow model. Modern video compression standards such as MPEG-4 [1998] allow for such arbitrary encoding of regions of support. However, finding these regions reliably is a difficult computational problem. For example, in the simple case of translational image motion, the outlines of the objects are the only regions where the image motion is discontinuous. Thus, an accurate estimate of the image motion is sufficient to find motion discontinuities. However, how can one extract motion discontinuities (based on image motion) if the outlines are a priori necessary for a good image motion estimate?

One can address this dilemma using an optimization process that estimates optical flow and the regions of support in parallel, but in the context of strong mutual recurrence so that their estimates are recurrently refined until they converge to an optimal solution. Given the flow model, this process will provide the estimate of the optical flow and thus the parameters of the model as well as the region boundaries for which the particular parameters hold. Such computational interplay also reflects the typical integrative and differential interactions

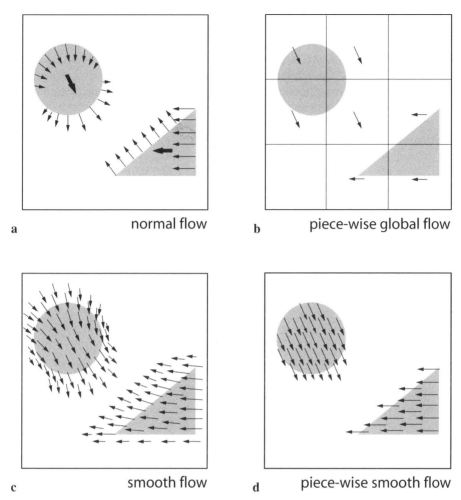

a   **normal flow**   b   **piece-wise global flow**

c   **smooth flow**   d   **piece-wise smooth flow**

Figure 2.6 *Motion integration.*
Two rigid objects are undergoing translational motion while the observer is not moving.
(a) Normal flow is only defined at brightness edges and is affected by the aperture problem.
(b) Block-based motion estimation provides a sparse representation (one motion vector per
block), but the image motion estimate is affected by the fixed partition scheme. (c) Smooth
integration kernels lead to appropriate estimates but do not preserve motion discontinuities.
(d) Ideally, a correct estimation for the optical flow field assumes two independent translational models, restricted to disconnected regions that represent the extents of each object.
The boundaries of these areas represent locations of discontinuous motion.

found in human psychophysical experiments known as *motion capture* and *motion contrast*
respectively [Braddick 1993]. For non-transparent motion, *motion segmentation* is the result
of the optimization process, in which the scene is segmented into regions of common motion
sources and each region is labeled with the model parameter (e.g. the optical flow vector $\vec{v}$).

Motion discontinuities

A further hint that motion discontinuities play an important role in visual motion perception is found in biological systems. A general property of biological nervous systems is their ability to discard redundant information as early as possible in the processing stream, and so reduce the huge input information flow. For example, antagonistic center-surround-type receptive fields transmit preferentially spatio-temporal discontinuities in visual feature space while regions of uniform visual features are hardly encoded. This encoding principle is well known for the peripheral visual nervous system such as the retina that receives direct environmental input. However, it also holds within cortical visual areas such as V1 (e.g. [Hubel and Wiesel 1962]) and higher motion areas like MT/V5 in primates [Allman et al. 1985, Bradley and Andersen 1998]. Furthermore, physiological studies provide evidence that motion discontinuities are indeed separately encoded in the early visual cortex of primates. Lamme et al. [1993] demonstrated in the awake behaving macaque monkey that motion boundary signals are present as early as in V1. In addition, studies on human subjects using functional magnet resonance imaging (fMRI) conclude that motion discontinuities are represented by retinotopically arranged regions of increased neural activity, spreading from V1 up to MT/V5 [Reppas et al. 1997]. These findings at least suggest that information about motion discontinuities seems useful even for cortical areas that are not considered to be motion specific.

Such data support the notion that motion discontinuities play a significant role already in the very early stages of visual processing. The findings suggest that the detection of motion discontinuities is vital. Further evidence proposes that the extraction of motion discontinuities is significantly modulated and gated by recurrent feedback from higher level areas. A study by Hupe et al. [1998] showed that the ability of neurons in V1 and V2 to discriminate figure and background is strongly enhanced by recurrent feedback from area MT. Clearly, such feedback only makes sense if it has a beneficial effect on the local visual processing. It seems rather likely that such recurrent, top-down connections do play a major role in the computation of motion discontinuities that help to refine the estimation of image motion.

2.5.3   Perceptual bias

Available prior information about the motion field can also help to resolve ambiguities in the image motion. For example, the knowledge that the two frames with the transparent striped pattern in Figure 2.3 can only move horizontally immediately resolves the ambiguity and reveals the transparent nature of the motion. Consequently, the interpretation of the image motion is biased by this *a priori* knowledge. Often, prior information is given in a less rigid but more statistical form like "usually, the frames tend to move horizontally...". Then, if many solutions are equally plausible, the estimate of image motion would be most likely correct if choosing the estimate that is closest to "move horizontally". Including prior information is very common in estimation theory (Bayesian estimator) but also in applied signal processing (Kalman filter [Kalman 1960]). Basically, the more unreliable and ambiguous the observation is, the more the system should rely on prior information. Noise is another source that leads to unreliable observations. So far noise has not been considered in the task of visual motion estimation. This book, however, addresses visual motion estimation in the context of physical autonomous systems behaving in a physical environment, where noise is self-evident. Part of the noise is internal, originating from the devices and circuits in the motion perception system. The other part is external noise in the visual environment.

From a computational point of view, using prior information can make the estimation of image motion well-conditioned for ambiguous visual input and under noisy conditions (well-posed in the noise-free case).[4]

2.6   Outline for a Visual Motion Perception System

A complete model for the estimation of optical flow typically combines a matching method and an appropriate flow model. The computational task of estimating optical flow then consists of finding the solution that best matches the model given the visual information. This computation can be trivial such as in the case of normal flow estimation, or it can become a very demanding optimization problem as in the recurrent computation of the optical flow and the appropriate motion discontinuities. In this case, the estimation of the optical flow field, given the visual input and the regions of support, usually involves a maximum search problem where the computational demands increase polynomially in the size of the regions of support N. On the other hand, the selection of the regions of support and thus the segmentation process remains a typical combinatorial problem where the number of possible solutions are exponential in N. Real-time processing requirements and limited computational resources of autonomous behaving systems significantly constrain possible optimization methods and encourage efficient solutions.

So far, I have briefly outlined what it means to estimate visual motion and what type of problems arise. One aspect which has not been stressed sufficiently is that the visual motion system is thought of being only part of a much larger system. Being part of an autonomous behaving system such as a goal-keeper, fly, or robot puts additional constraints on the task of visual motion estimation. As discussed briefly, the computation must be robust to noise. The estimates of image motion should degrade gracefully with increasing amounts of noise. Also, the visual motion system should provide for any possible visual input a reasonable output that the rest of the system can rely on. It should not require supervision to detect input domains for which its function is not defined. And finally, it is advantageous when higher processing areas can influence the visual motion system to improve its estimate of image motion. This top-down modulation will be discussed in later chapters of the book.

Now, following the discussion above I want to return to Figure 1.1 and replace the unspecified transformation layer in that figure with a slightly more detailed processing diagram. Figure 2.7 outlines a potential visual motion system that is part of an efficient, autonomously behaving agent. The system contains or has access to an imager that provides the visual information. Furthermore, it performs the optimization task of finding the optimal estimate of image motion, given the data from the matching stage, and the flow model that includes prior information about the expected image motion. The system provides an estimate of image motion to other, higher level areas that are more cognitive. In recurrence, it can receive top-down input to improve the image motion estimate by adjusting the flow model.

---

[4]See Appendix A.2.

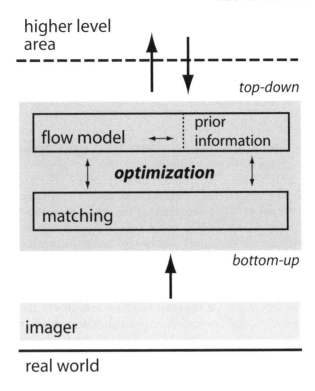

Figure 2.7 *Outline of a visual motion perception system.*
The system includes a matching model and a flow model that uses prior information. Optimization is needed to find the optimal optical flow estimate given the visual input and the model. Recurrent top-down input can help to refine the motion estimate by adjusting the model parameters (e.g. priors, region of support). This top-down input also provides the means to incorporate attention.

2.7   Review of aVLSI Implementations

This chapter ends with a review of aVLSI implementations of visual motion systems. The listing is complete to my best knowledge, but includes only circuits and systems that were actually built and of which data have been published. Future implementations will be included in an updated list that can be found on the book's webpage (http://wiley.com/go/analog).

Several approaches apply *explicit matching in the time domain*: they compute the time-of-travel for a feature passing from one detector to its neighbor. In [Sarpeshkar et al. 1993] and later elaborated in [Kramer et al. 1997], the authors propose two circuits in which the matching features are temporal edges. In one of their circuits a temporal brightness edge triggers a pulse that decays logarithmically in time until a second pulse occurs indicating the arrival of the edge at the neighboring unit. The decayed voltage is sampled and represents the logarithmically encoded local image velocity. The 1-D array exhibits accurate velocity estimation over many orders of magnitude. In the second scheme, the temporal edge elicits a pulse of fixed amplitude and length at the measuring detector as well as at its neighbor. The overlapping pulse width is an inversely linear measure for image velocity. The operational range is limited in this scheme. Either small velocities are not detectable because the pulses become non-overlapping or the resolution for high velocities decreases. An additional variation [Kramer 1996] encodes image velocity as inversely proportional to the length of a binary pulse. Local motion circuits of this kind were used successfully for the localization of the focus-of-expansion [Indiveri et al. 1996, Higgins et al. 1999] or the detection of 1-D motion discontinuities [Kramer et al. 1996]. Similar implementations have been reported by Higgins and Koch [1997].

Another aVLSI implementation of explicit matching in the time domain uses spatial edges as matching features [Etienne-Cummings et al. 1993, Etienne-Cummings et al. 1997]. Spatial edge detection is achieved by approximating a difference-of-Gaussian (DOG) operation. Local brightness values are compared to the smoothed image provided by a resistive mesh using well implants or polysilicon lines as resistive elements [Etienne-Cummings 1994]. The presence and the location of spatial edges are coded as binary signals that are obtained by thresholding the edge-enhanced image. Velocity is measured as the inverse time-of-travel for the binary edge signals. A similar approach has been presented recently where brightness edges are extracted as the thresholded brightness ratio between neighboring pixels as opposed to the DOG filter of the previous approach [Yamada and Soga 2003]. Miller and Barrows suggested a modification of that scheme by extracting multiple types of local features to obtain a more robust estimate. Local features are extracted with simple binary filter kernels [Miller and Borrows 1999].

Horiuchi et al. [1991] transformed the velocity estimation into a spatial correlation task, inspired by the mechanism of sound localization in the barn owl. The occurrence of a feature at a detector and its neighbor triggers two pulses that travel along oppositely directed delay lines. Correlation circuits between these lines detect where pulses pass each other. Active correlation circuits are detected by a winner-takes-all circuit, and provide a spatial encoding of the observed image velocity. Slow speeds in either direction will be detected at the ends of the correlation array, whereas fast-speed winners lie close to the middle.

Moini et al. [1997] discuss a 1-D motion chip that emulates the typical insect visual motion detection system. The implementation simplifies the biological system insofar as it quantizes the temporal changes in the visual input signal. The circuit uses the quantized temporal changes as labeled events to apply template matching in the time domain in order to extract visual motion information. Matching is performed off-chip to permit the use of different template models with different complexity levels. Thus, the chip provides only the frontend of a visual motion system. Special attention was paid to a multiplicative noise cancellation mechanism that reduces the sensitivity of the chip to flicker noise.

One of the few examples of a successful commercial application of integrated motion-sensing circuits is Logitech's Trackman Marble,[5] a trackball pointing device for personal computers [Arreguit et al. 1996]. So far, several million units have been sold world-wide. The core of the trackball is a 2-D optical sensor array of 75 motion processing units that uses spatio-temporal sampling to extract the global motion of the trackball. The implementation is different from the previous examples because it requires precisely controlled and pulsed lighting conditions, as well as a specific random dot pattern stimulus that is directly

---

[5] www.logitech.com

imprinted onto the trackball itself. Obviously, these well-controlled visual conditions substantially simplify the motion extraction task. Each pixel consists of a simple edge detector
that compares the differences in reported brightness between neighboring photoreceptors
to a given threshold. The global motion of the trackball in between two time-steps is then
directly computed by dividing the number of edges that moved in each of two orthogonal
directions by the total number of edges present. The higher the total number of edges seen
by the sensor, the higher the resolution of the motion estimate. Using a particular random
dot pattern, the spatial average of the number of edges can be well controlled and kept
rather constant. The temporal sampling has to be fast enough that the maximal on-chip displacement is guaranteed to be smaller than one pixel-spacing, otherwise temporal aliasing
will occur.

In general, the circuits presented above are compact, show a robust behavior, and
work over a wide range of image motion. However, there is a fundamental limitation of
algorithms that rely on measuring the time-of-travel of features across the image plane:
they can only report local image motion after the moving feature was actually detected
at one local motion detector and reappeared at one of its neighbors. These algorithms are
undetermined in situations where no motion is present. Zero motion is by definition not
part of their operation range because it implies an infinitely long travel time. In practice, a
maximal time interval is defined (either explicitly or implicitly) within which the feature is
expected to travel between two local motion detectors. This is equivalent to a threshold for
detecting low speeds. Every visual input that does not elicit a response is assumed to have
zero motion. The algorithms cannot distinguish between, for example, a visual scene that
does not contain any features and a visual scene that shows clear brightness patterns but
does not move. Both will not elicit a response, and thus are assumed to have zero motion,
yet there is a clear qualitative difference between the two scenes.

The next class of implementations applies *gradient-based* matching models. Tanner
and Mead [1986] proposed a chip which was one of the earliest attempts at building
analog integrated motion circuits. Its 8×8 array consists of identical motion units that
collectively compute a common global motion estimate. The two components of the global
motion vector are provided to each unit as two voltages on global wires. Each unit adjusts
this global estimate according to its locally measured image gradients such that the sum
of the squared errors of the constraint Equation (2.4) is minimal across the whole array.
Thus, the chip implements the computational approach proposed by Lucas and Kanade
[1981] with uniform weighting and a window size equivalent to the whole imaging array.
In principle, the chip is able to solve the aperture problem for singular motion and so
computes real 2-D motion. Apparently, however, the chip never showed a robust behavior
under real-world conditions.[6] The feedback computation of the local error was difficult to
control and sensitive to mismatch and noise in the brightness gradient computation. A first
attempt to improve the implementation also did not achieve a considerably more robust
behavior [Moore and Koch 1991].

A gradient-based 1-D implementation has been reported by Deutschmann and Koch
[1998a]. In one spatial dimension, normal flow can be simply computed as $v = -\partial E_t / \partial E_x$.
To avoid division by zero, a small constant current is added to the denominator. The use
of wide-linear-range circuits allows the chip to respond over two orders of magnitude of

---

[6]Personal communications, T. Delbrück and J. Harris and C. Koch

velocity range. A recent approach proposes an extension to two spatial dimensions [Mehta and Etienne-Cummings 2004]. Unlike most other implementations discussed here, this chip applies a sequential processing scheme where the imaging and the motion processing circuitry are physically separated. Motion processing is done off-array. The advantage is a high-fill factor, high-resolution image sampling, whereas the disadvantage is the temporally sampled processing. By computing $u = -\partial E_t / \partial E_x$ and $v = -\partial E_t / \partial E_y$, the estimation of normal flow is only approximated (see Equation (2.5)) by basically computing 2×1-D normal flow on a 2-D array. The approximation particularly deviates from a "true" normal flow estimate at image locations where there are large differences between the spatial brightness gradients in the x and y directions.

Compact gradient-based motion sensors were also presented [Moore and Koch 1991, Horiuchi et al. 1994] and later in two spatial dimensions [Deutschmann and Koch 1998b] where the resulting motion signal is the product of the spatial and temporal brightness gradients. Such a motion signal obviously strongly depends on the contrast and spatial frequency of the visual scene. Nevertheless, these compact implementations provide at least the correct direction of normal flow. This is sufficient for visual motor tasks such as saccadic and smooth pursuit tracking [Horiuchi et al. 1994].

A third class comprises implementations that closely model *correlation-based methods* observed in biological systems. There is a series of implementations based on the Reichardt scheme [Hassenstein and Reichardt 1956]. An early representative [Andreou and Strohbehn 1990] consists of a 1-D array of motion detectors where multiplication between the spatially filtered visual input and its delayed version from a neighboring detector unit serves as the non-linear correlation function. The correlation signals from local detectors are averaged to provide a global motion estimate. The first 2-D implementation uses unidirectional delay lines as temporally tuned filters for moving edges [Delbrück 1993c]. The delay lines are oriented in three spatial directions within a hexagonal array of correlation units. Photoreceptors report brightness changes and couple their output into the delay lines where they propagate with a characteristic velocity in one of three directions. Whenever the velocity of a moving brightness edge matches the propagation speed along a particular delay line, then the brightness signal on the delay line is reinforced. The circuit performs temporal integration of the motion signal along the orientations of the delay lines. Local motion information is blurred by means of the temporal integration. Nevertheless, this approach utilizes a basic form of collective computation, since the motion estimate is the result of spatio-temporal integration along the elements of each delay line.

Jiang and Wu [1999] reported a correlation-based approach where motion is reported if the time-of-travel of extracted brightness edges between neighboring motion units matches a preset delay time. Since the tuning is very narrow, the sensor is fairly limited and cannot respond to arbitrary visual velocities without being continuously retuned.

Benson and Delbrück [1992] describe a direction-selective retina circuit that is inspired by the mechanism responsible for direction selectivity in the rabbit retina [Barlow and Levick 1965]. This circuit uses a similar approach as Kramer [1996]: a temporal edge detected at a local motion unit elicits a pulse which is suppressed by inhibitory input from the neighboring motion units in the preferred direction. Thus the length of the output pulse is inversely proportional to the speed. In the null direction, inhibition suppresses the output of the neighboring detector. The chip has an array of 41×47 direction selective units, all of which have the same directional tuning.

Recently, the fly's visual motion system attracted some attention from neuromorphic engineers. One-dimensional silicon models of the fly's elementary motion detector have been described that show similar tuning characteristics to physiological data from the insects [Harrison and Koch 1998, Liu 2000]. The estimate for visual motion is computed by multiplying the temporal high-pass-filtered signal of the photoreceptor of a motion detector with the low-pass-filtered signal of its neighbor. The fixed time constants of the low-pass and high-pass filters make the detector narrowly tuned to particular spatial and temporal frequencies of the input stimulus. Liu [2000] proposes a silicon model of the spatial aggregation of the local motion responses observed in the fly that leads to a global motion output, which is relatively insensitive to the contrast and size of the visual stimulus. Adaptation of the time constants of the low- and high-pass filters is also known to take place in the fly's visual motion system such that the sensitivity to changes in the perceived motion remains high. Such adaptation mechanisms have been implemented and show successful replication of physiological data [Liu 1996].

These existing correlation-based implementations estimate visual motion along only one spatial dimension. They report visual motion but not velocity explicitly. Motion signals from a correlation detector are inherently dependent on contrast, spatial, and temporal frequency. Correlation-based methods are part of the more broadly defined class of *motion energy-based* approaches. The estimation of 2-D motion requires large ensembles of spatio-temporally oriented filters. Furthermore, the extraction of apparent image velocity requires additional processes such as integration [Heeger 1987b] or normalization [Adelson and Bergen 1985, Simoncelli and Heeger 1998]. There have been attempts to realize a full motion energy-based system using special purpose hybrid hardware [Etienne-Cummings et al. 1996, Etienne-Cummings et al. 1999]. This system only approximates 2-D spatial filtering. A single motion unit with filters of combinations of only three spatial and three temporal scales requires several hundred neuronal units with thousands of synapses.

Little is known about implementing aVLSI *motion segmentation* systems. Although several implementations have been reported that successfully demonstrate segmentation in 1-D feature space [Harris and Koch 1989, Harris 1991, Liu and Harris 1992], there is only one known attempt by Kramer et al. [1996] to apply resistive fuse circuits for motion segmentation. This implementation contains a 1-D array of local motion elements as described in Kramer et al. [1995]. Bump circuits [Delbrück 1993a] compare the output of two adjacent motion elements and control the conductance in a diffusion network accordingly. In regions of common motion, the output of the individual elements is averaged while regions of different motion are separated. The 2-D array by Lei and Chiueh [2002] applied non-isotropic diffusion to approximate motion segmentation. Motion integration, thus the connection strength between neighboring motion units, is thereby modulated according to the brightness gradients in the image. The problem with this approach is that brightness gradients usually occur not only at motion discontinuities but also within object textures.

Table 2.1 summarizes the review of existing aVLSI visual motion systems. The majority of the systems estimates visual motion only along single spatial dimensions, so avoiding the aperture problem and therefore complex integration schemes. Some of these circuits can be very compactly integrated. However, to estimate visual motion in two spatial dimensions, information about the 2-D structure of the image scene must be available. Gradient-based

Table 2.1 *Overview of existing aVLSI motion systems.*
Those aVLSI visual motion circuits (systems) are listed that were implemented and
where chip performance was reported. A continuously updated version of this listing
can be found on-line (http://wiley.com/go/analog).

Authors	Motion resolution	Motion dimension	Explicit velocity	Motion integration
Optical flow estimation				
Feature matching				
[Horiuchi et al. 1991]	local	1-D	yes	no
[Etienne-Cummings et al. 1993]	local	2 × 1-D	yes	no
[Sarpeshkar et al. 1993, Kramer et al. 1997]	local	1-D	no	no
[Arreguit et al. 1996]	global	2-D	yes	spatial avg.
[Kramer 1996]	local	1-D	yes	no
[Higgins and Koch 1997]	local	1-D	yes	no
[Moini et al. 1997]	local	1-D	yes[a]	no
[Miller and Borrows 1999]	local	1-D	yes	no
[Yamada and Soga 2003]	local	1-D	yes	no
Gradient-based				
[Tanner and Mead 1986]	global	2-D	yes	IOC
[Moore and Koch 1991]	local	1-D	no	no
[Horiuchi et al. 1994]	local	1-D	no	no
[Deutschmann and Koch 1998a]	local	1-D	yes	no
[Deutschmann and Koch 1998b]	local	2-D	no	no
[Mehta and Etienne-Cummings 2004]	local	2 × 1-D	yes	no
The smooth optical flow chip (page 127) [Stocker and Douglas 1999 (early version), Stocker 2006]	local...global	2-D	yes	IOC
Correlation-based				
[Andreou et al. 1991]	global	1-D	no	spatial avg.
[Benson and Delbrück 1992]	local	1-D	no	no
[Delbrück 1993c]	local	3 × 1-D	no	temporal avg.
[Harrison and Koch 1998]	local	1-D	no	no
[Jiang and Wu 1999]	local	2-D	no	no
[Liu 2000]	global	1-D	no	spatial avg.

(*continued overleaf*)

Table 2.1   (*continued*)

Authors	Motion resolution	Motion dimension	Explicit velocity	Motion integration
Motion energy models				
[Etienne-Cummings et al. 1999]	local	2 × 1-D	yes	no
Motion segmentation				
[Kramer et al. 1996]	local	1-D	yes	no
[Lei and Chiueh 2002]	local	2-D	yes	non-isotropic
The motion segmentation chip (page 157) [Stocker 2002, Stocker 2004]	local...global	2-D	yes	line-process/IOC
The motion selection chip (page 167) [Stocker and Douglas 2004]	local...global	2-D	yes	non-isotropic/IOC

[a]Not reported, but possible with off-chip template matching (see description on page 25).

methods have the advantage that the spatial gradients are relatively easy to compute and are a compact representation of spatial orientation. Explicit matching methods would require a feature extraction stage that accounts for 2-D features. Implementations for the extraction of brightness corners have been reported [Pesavento and Koch 1999] but they require rather expanded circuitry. Motion energy-based systems are probably the method of choice but turn out to be expensive to implement even in a simplified manner.

3

Optimization Networks

The main computational task of the visual motion system, as outlined in the previous chapter (Figure 2.7), is optimization. Ideally, the system finds the optimal visual motion estimate with respect to the applied matching and flow models and the given visual information. This chapter introduces *constraint satisfaction* as a framework for formulating such optimization problems, and shows how to design recurrent networks that can establish appropriate solutions. Two simple examples serve to illustrate the methodology, and will be revisited in the context of computational network architectures described in later chapters.

3.1   Associative Memory and Optimization

Since the ground-breaking work of McCulloch and Pitts [1943] many different mathematical descriptions of computation in neural networks have been proposed. Amongst these, models of how neural networks can store and retrieve information were of major interest. Beginning in the early 1970s, the concept of *associative content-addressable memory* became popular. Network models of associative memory are able to learn different prototype patterns and, when presented with a new pattern, to associate it to the stored prototype that is most similar. While many key ideas of associative memory were developed much earlier [Cragg and Temperley 1955, Little and Shaw 1975, Grossberg 1978], it was John Hopfield who established a network model that was clearly defined and had a strong physical interpretation [Hopfield 1982, Hopfield 1984]. He showed that in an associative memory, each pattern can be represented by the locally stable states (attractors) of a dynamical network of essentially binary units. Each stable state represents a local minimum of a cost function.[1] The exact form of the cost function is defined by the architecture (weights and connectivity) of the network. The connectivity of these Hopfield networks is generic and usually isotropic, for example fully connected. The weights of the connections are typically adjusted by a local learning rule to store the desired prototype patterns (e.g. Hebbian learning [Hebb

---

[1]Using the expression *cost function* instead of *energy* avoids potential confusion with the physical meaning of energy.

1949]). After the learning phase each prototype pattern is associated with one distinct local attractor of the network as long as the memory capacitance is not exceeded. Presented with a new input pattern, the network state converges asymptotically to an attractor that represents the learned prototype pattern that is closest to the input pattern.[2]

Hopfield's notion that the dynamics and thus the activity in such networks are explained entirely by the search for the local minima of a cost function immediately reveals the potential for solving optimization problems with such attractor networks: if one could formulate a given optimization problem as the minimization of a cost function, then an appropriately designed network would find a local solution to that problem. Hopfield himself (in collaboration with David Tank) demonstrated the potential of the approach by describing a network solution for the well-known *traveling salesman problem (TSP)* [Hopfield and Tank 1985]. The TSP is a computationally hard optimization problem. It considers a salesman who must choose the shortest overall path to visit a number of cities only once each. This is a typical combinatorial problem in which the number of possible solutions grows exponentially with the number of cities. Their proposed network could find (near-)optimal solutions within only a few characteristic time constants. Although the particular network solution to the TSP turned out not to scale very well beyond 30 cities [Wilson and Pawley 1988, Gee and Prager 1995] it clearly was an impressive demonstration of the potential of the approach. Meanwhile, a variety of optimization problems including linear and quadratic programming [Tank and Hopfield 1986], and their generic network solutions, have been addressed [Liang and Wang 2000]. Readers who are interested in a more complete treatment of the subject should also read the excellent textbooks by Hertz et al. [1991] and Cichocki and Unbehauen [1993].

What makes the network solution to the TSP and other, similar combinatorial problems so appealing is their efficiency. Unlike a search algorithm the network does not consider each possible solution sequentially. Rather, it evaluates all possible solutions simultaneously during the solution-finding process, not establishing a definite solution until it finally converges. Starting from an initial state the network could evolve into many possible solutions for a given input. But its architecture is such that it encourages convergence only toward the desired solutions. Because the solution is represented as the ensemble state of all individual units in the network, each small change in the local state of a unit immediately effects all the other units, making the solution the result of a *parallel collective* computational effort. It is obvious that the true power of such optimization networks can only be exploited when these networks are implemented as a parallel computational architecture, and not only simulated on a sequential digital computer. This is the key motivation for the aVLSI implementations presented later in this book.

3.2   Constraint Satisfaction Problems

Recall that the estimation of visual motion is an optimization problem that seeks the optimal optical flow estimate given the visual observation, the matching and flow model, and, possibly, additional input from higher cognitive levels. These inputs and models impose

---

[2]The metric is thereby defined by the network architecture. In a typical linear network the distance between a stored pattern $\vec{x}_0$ and the input pattern $\vec{x}$ is measured as the Hamming distance (binary patterns) or more generally as the Euclidean distance $||\vec{x} - \vec{x}_0||^2$ (2-norm).

individual constraints on the solution. Thus finding the optimal solution is called a constraint satisfaction or constraint optimization problem. Its mathematical formulation and the computational approach chosen depend on the constraints imposed. In the case of estimating optical flow, the constraints imposed by the model (e.g. constant brightness) are usually only approximately true with respect to the visual observation. Thus, it is sensible to apply an optimization approach where a solution does not require the constraints to be exactly fulfilled in order to be valid. Rather, the optimal solution should best possibly satisfy the constraints. Such an approach can be formulated as the unconstrained optimization problem[3]

$$\text{minimize} \quad H(\vec{x}) = \sum_n \alpha_n C_n(\vec{x}) \quad \text{subject to } \vec{x} \in \mathbb{R}^m, \tag{3.1}$$

where each $C_n(\vec{x})$ is a measure of how strongly the possible solution $\vec{x}$ violates the n-th constraint, and α_n determines its importance or weight. Each constraint reflects partial properties of a model that is characterized by the cost function $H(\vec{x})$. The weighting parameters provide a limited way of programming so that the model can be adapted to solve different optimization problems. Also, these parameters can be controlled from another top-down process, providing the necessary dimensional reduction used in the control structure of complex systems. This will become important in the discussion of recurrent systems.

Constrained optimization problems [Fletcher 1981], which require the exact fulfillment of some of their constraints, can also be formulated in the unconstrained form (3.1) using Lagrange multipliers. However, they might not be well described that way. For example, Hopfield and Tank's network solution for the TSP does not scale well because of this reason. The *syntactic constraints* of the problem (every city once, only one city at a time) are constraints that must precisely hold for any valid solution (tour). Hopfield and Tank did formulate the TSP as unconstrained optimization problem (3.1) in which the syntactic constraints compete with the other constraints of the problem related to the position and distance of the cities. With an increasing number of cities, the subset of syntactically valid solutions compared to the total number of possible solutions decreases exponentially. As a result, the ratio of valid to the total number of solutions found drops to a inefficiently low value for a TSP with more than 30 cities. A different form of the syntactic constraints can help to improve on the efficiency of the original network solution (see e.g. [Cichocki and Unbehauen 1993] for a review).

3.3 Winner-takes-all Networks

The first example that will illustrate how to derive network solutions for unconstrained optimization problems is a simple competitive task: given a discrete set of positive input values $E_1, \ldots, E_N \subset \mathbb{R}^+$, find an index *max* such that

$$E_{max} \geq E_j \ \forall \ j \neq max. \tag{3.2}$$

This describes the well-known winner-takes-all (WTA) operation. In order to derive a network solution, the problem is reformulated as a mapping operation: find a network architecture of N computational units that maps a given input vector $\vec{E} = (E_1, \ldots, E_N)$

---

[3]The literature typically calls the optimization problem (3.1) *unconstrained* because it enforces no additional constraints on the solution other than minimizing the cost function [Fletcher 1980, Fletcher 1981].

to a binary output vector $\vec{V} = (V_1, \ldots, V_N)$ such that $V_{max} = 1$ and $V_{j \neq max} = 0$. This mapping operation can be expressed as an unconstrained optimization problem by specifying a number of suitable constraints. The following cost function combines three constraints that sufficiently define the WTA problem:

$$H(\vec{V}) = \underbrace{\frac{\alpha}{2} \sum_i \sum_{j \neq i} V_i V_j}_{\text{sparse activity}} + \underbrace{\frac{\beta}{2} \left(\sum_i V_i - 1 \right)^2}_{\text{limited total activity}} - \underbrace{\gamma \sum_i V_i E_i}_{\text{biggest input wins}} \quad . \tag{3.3}$$

The relative strengths of the constraints are given by the weighting parameters α, β, and γ. The first and the second constraint promote all the states where only one output unit is active and its activity level $V_{max} = 1$. These are the syntactic constraints that ensure that the network performs a decision in every case, whether it is the right one or not. The third constraint finally relates the input to the output such that it will be most negative if the winning unit is the one that receives largest input. For finite input values, the cost function is bounded from above and below where the lower bound is equal to $H_{min} = -\gamma E_{max}$. If the chosen constraints are appropriate, then the solution for the WTA problem (3.2) is equivalent to the solution of the unconstrained optimization problem: given the input $\vec{E}$, find the output $\vec{V}$ such that $H(\vec{V})$ is minimal.

With a cost function assigned, a local update rule is needed that can be applied repeatedly to each computational unit in the network such that from a given start condition the network converges to a state of minimal cost. So far, let us assume that the output of each unit can be only in one of two states [0, 1]. A sensible rule dictates changing the output state V_i of unit i only if this lowers the total cost. The change in cost for a transition $V_i \rightarrow V_i'$ of a single unit's output can be written as

$$\Delta H_{V_i \rightarrow V_i'} = H(\vec{V}') - H(\vec{V}). \tag{3.4}$$

Using a finite difference approximation $\Delta H_{V_i \rightarrow V_i'}$ can be computed from Equation (3.3) as

$$\Delta H_{V_i \rightarrow V_i'} = \Delta V_i \left(\alpha \sum_{j \neq i} V_j + \beta \left(\sum_i V_i - 1 \right) - \gamma E_i \right). \tag{3.5}$$

The update rule can now be formulated in terms of positive ($\Delta V_i = 1$) and negative ($\Delta V_i = -1$) transitions:

$$V_i' \Rightarrow \begin{cases} \text{apply transition} & \text{if } \Delta H_{V_i \rightarrow V_i'} < 0 \\ \text{no transition} & \text{otherwise.} \end{cases} \tag{3.6}$$

A WTA network with the above dynamics does not consider the presence of any thermal noise. A stochastic formulation of the dynamics at a given finite temperature would be more realistic but requires methods of statistical mechanics to describe the network behavior and find valid solutions.

Instead, a different approach is followed here. The network is modified so that the units are permitted to have continuous valued outputs. The response of a single unit is now described by a continuous activation function $g(u) : u \rightarrow V$, where u represents the internal

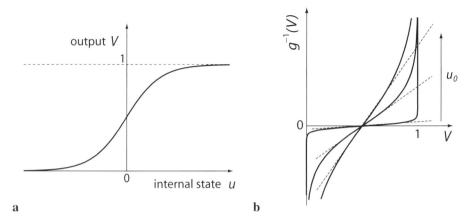

Figure 3.1 *Activation function.*
(a) A typical sigmoidal activation function is shown. (b) The inverse of the sigmoidal activation function (3.7) is plotted for different slope parameters u_0. The total integral under each curve from zero to a given value V represents the activation cost needed to keep that unit's output at V. It decreases with decreasing u_0.

state of the unit and V its output. A typical sigmoidal activation function, for example, is

$$g(u) = \frac{1}{2}(\tanh(u/u_0) + 1),\qquad(3.7)$$

and is shown in Figure 3.1. The parameter $1/u_0$ characterizes the maximal response gain of the unit. In the limiting case where $1/u_0 \to \infty$ the output of the units becomes approximately binary. The activation function, however, does not have to be sigmoidal. The WTA network requires only that the function is differentiable, monotonically increasing, and bounded from below. In particular, the output range of the units is not necessarily limited to the unity interval because the second constraint limits the total output activity in the network anyway. Thus, piece-wise linear or linear-threshold units can be also used, and can be treated within the same analytical framework. The only relevant parameter is the activation gain $1/u_0$.

With continuous units, the constraints imposed on the WTA network remain the same, although the output of the winning unit is not guaranteed to be exactly one. Also the cost function (3.3) will not exactly remain the same. It must include an additional term that represents the total activation cost needed to keep the unit active. The activation cost can be considered as an additional constraint imposed by the computational machinery rather than by the problem itself. Thus, the cost function (3.3) is modified to

$$\tilde{H}(\vec{V}) = \frac{\alpha}{2}\sum_i\sum_{j\neq i}V_iV_j + \frac{\beta}{2}\left(\sum_i V_i - 1\right)^2 - \gamma\sum_i V_iE_i + \underbrace{1/R\sum_i\int_{1/2}^{V_i}g^{-1}(\xi)\,\mathrm{d}\xi}_{\text{activation cost}},$$

$$(3.8)$$

where $1/R$ is the weighting parameter for the activation cost.[4] Figure 3.1b shows that the integral over the inverse sigmoidal activation function, thus the activation cost for a single unit, decreases with increasing activation gain. In the high-gain limit ($u_0 \to 0$) when the units essentially become binary, the cost function (3.8) approaches the previous form (3.3) as expected.

$\tilde{H}(\vec{V})$ is sufficiently regular and at least twice continuously differentiable because $\vec{V}$ is continuous. Thus, *gradient descent* on the cost function (3.8) can be used as the new local update strategy for the units in the network. Following gradient descent, the output V_i of each unit should change proportionally to the negative partial gradient of the cost function,

$$\dot{V}_i \propto -\frac{\partial \tilde{H}}{\partial V_i}, \tag{3.9}$$

until a steady state of local minimal cost is reached where $\partial \tilde{H}/\partial V_i = 0 \ \forall \ i$. Since V_i is monotonic in u_i, gradient descent defines the following dynamics on the internal state of the units:

$$\dot{u}_i = -\frac{1}{C} \frac{\partial \tilde{H}}{\partial V_i}$$

$$= -\frac{1}{C} \left[\frac{u_i}{R} + \alpha \sum_{j \neq i} V_j + \beta \left(\sum_i V_i - 1 \right) - \gamma E_i \right], \tag{3.10}$$

where C defines the time constant. Yet, it has to be examined whether the proposed dynamics guarantee a stable system that always converges to some asymptotically stable fixed point. To prove this, it is sufficient to show that for arbitrary starting conditions, the cost function (3.8) always decreases when following the dynamics (3.10) [Hopfield 1984]. Differentiating $\tilde{H}(\vec{V})$ with respect to time leads to

$$\frac{d\tilde{H}}{dt} = \sum_i \frac{\partial \tilde{H}}{\partial V_i} \dot{V}_i$$

$$= \alpha \sum_i \sum_{j \neq i} V_j \dot{V}_i + \beta \left(\sum_i V_i - 1 \right) \sum_i \dot{V}_i - \gamma \sum_i \dot{V}_i E_i$$

$$+ 1/R \sum_i g^{-1}(V_i) \dot{V}_i$$

$$= \sum_i \dot{V}_i \left[\sum_j (\alpha + \beta) V_j - \alpha V_i - \beta - \gamma E_i + \frac{u_i}{R} \right]$$

$$= -C \sum_i \dot{V}_i \dot{u}_i \qquad \text{(substitution with (3.10))}$$

$$= -C \sum_i (\dot{u}_i)^2 g'(u_i) \ \leq 0. \tag{3.11}$$

---

[4]In the context of an equivalent electronic network implementation, the parameter $1/R$ can be interpreted as the non-zero leak conductance of each unit.

Thus, $\tilde{H}(\vec{V})$ is never increasing, and because it is bounded from below it will always converge to an asymptotically stable fixed point. Therefore, $\tilde{H}(\vec{V})$ is a Lyapunov function of the system. This is true for all differentiable activation functions that are monotonically increasing, that is $g'(u) \geq 0$.

3.3.1   Network architecture

So far, the WTA task is formulated as an unconstrained optimization problem with cost function (3.8). The gradient descent dynamics (3.10) describe a *network* of N simple computational units with a particular, non-linear activation function. They guarantee that the system has asymptotically stable states representing the wanted solutions. However, it is not clear yet what the architecture of such a network could be. The dynamics can be rewritten as

$$\dot{u}_i = -\frac{u_i}{\tau} \; + \; \underbrace{\alpha V_i}_{\text{self-excitation}} \; - \; \underbrace{(\alpha + \beta) \sum_i V_i}_{\text{global inhibition}} \; + \; \gamma \tilde{E}_i \qquad (3.12)$$

by adding a constant term to the network input ($\tilde{E}_i = E_i + \beta/\gamma$), and normalizing the weight parameters α, β, and γ by C. In this reformulated form, the dynamics (3.12) define a network that exhibits self-excitation and global inhibition [Grossberg 1978]. The dynamics also permit a physical interpretation describing, for example, the voltage dynamics at the capacitive node u_i of the simplified electrical circuit shown in Figure 3.2. Kirchhoff's current law requires the sum of the capacitive current ($\dot{u}_i C$) and the leak current (u_i/R) to equal the total input current to the circuit, which consists of the external input (γE_i), a recurrent component (αV_i), and the global inhibition current from all units ($(\alpha + \beta) \sum_i V_i$). A typical sigmoidal activation function $g(u)$ can be implemented with the aid of a transconductance amplifier [Hopfield 1984], but other functions are also possible.

There are basically two interpretations of the network connectivity. It can be thought of as a network of homogeneous units where each unit has inhibitory connections to all the other units and one excitatory connection to itself. Or alternatively, inhibition can be provided by a single additional inhibitory unit that sums the equally weighted output from all N units and feeds it back to each unit as a global inhibitory input [Kaski and Kohonen 1994]. Both cases are illustrated in Figure 3.3. The need for an additional unit and the loss of homogeneity in the second solution are compensated by the massive reduction of connectivity. Applying a global inhibitory unit reduces the connectivity to $3/(N + 1)$ of the fully connected architecture, a substantial reduction to a few percent for networks consisting in the order of hundreds of units. To be computationally equivalent, however, the time constant of the inhibitory unit must be small compared to the dynamics of the other units. Also, the activation function of the inhibitory unit is assumed to be linear. Other, for example sigmoidal, activation functions would alter the cost function (3.8) such that α and β become dependent on the absolute level of activity, yet the qualitative behavior of the network would remain the same.

A reduction of the connectivity level can be important in a physical implementation of such networks. Each connection requires a wire of some kind or another, but wires are not for free in the physical world. They do need space. Reducing the total wiring costs increases

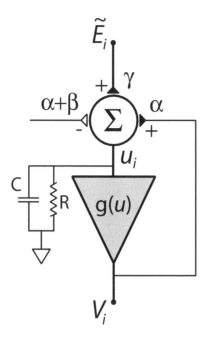

Figure 3.2 *WTA network unit with continuous output.*
A single unit of the WTA network receiving excitatory (full triangle) and inhibitory input (open triangle). The time constant of the unit is determined by the capacitance C and the leak conductance R.

the efficiency of the implementation by reducing its volume and weight. The wiring costs can be a crucial constraint in defining the layout of physical networks. This is certainly true for aVLSI implementations such as the ones discussed later in this book. It might also be true for biological implementation; for example, it has been argued that the particular structure of the human brain is the result of minimizing the total wiring length, assuming a given functional cortical connectivity pattern [Koulakov and Chklovskii 2001, Chklovskii 2004].

3.3.2 Global convergence and gain

Although the network has proven that it always converges to an asymptotically stable state, it is not clear that it is globally convergent: that is, always converges to the global minimum. In fact, a gradient descent update strategy always gets stuck in local minima and there is no deterministic method to avoid it. The only way to ensure global convergence is to make sure that there *are* no local minima.

Convexity of the cost function $\tilde{H}$ is a sufficient but strong condition for global convergence.[5] Obviously, there are non-convex functions that have no local minima. However, convexity is a useful condition because it is relatively easy to test: $\tilde{H}$ is guaranteed to be

---

[5]See Appendix A.3.

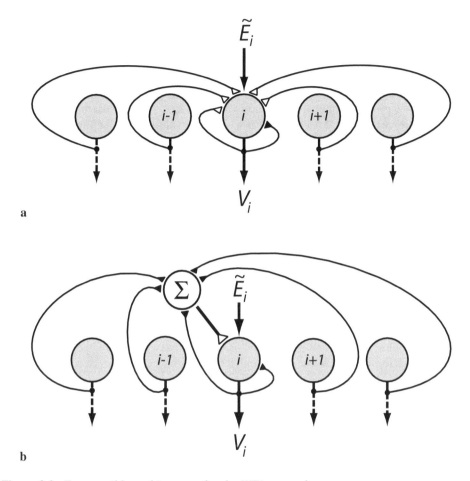

Figure 3.3 *Two possible architectures for the WTA network.*
(a) Homogeneous architecture where the connectivity for each unit is the same. Excitatory connections are indicated by full and inhibitory ones by open triangles. (b) WTA network with one special (global inhibitory) unit. In both architectures, input and inhibitory connections are only drawn for unit i.

convex if its Hessian $\mathbf{J}_{\tilde{H}}$ is positive semi-definite.[5] The Hessian $\mathbf{J}_{\tilde{H}}$ of (3.8) is symmetric and only depends on the weighting parameters (α, β, γ, and R). In order to be positive semi-definite the eigenvalues must be real and non-negative. There are two distinct eigenvalues for $\mathbf{J}_{\tilde{H}}$

$$\lambda_1 = \beta + \frac{u_0}{R} + (N-1)(\alpha + \beta) \quad \text{and} \quad \lambda_2 = -\alpha + \frac{u_0}{R}. \tag{3.13}$$

Assuming only positive weighting parameters, the first eigenvalue λ_1 is always positive whereas λ_2 is only non-negative for

$$\alpha \leq \frac{u_0}{R}. \tag{3.14}$$

The upper limit of condition (3.14) is called the *self-excitation gain limit*. Obviously, there is a trade-off between strong self-excitation (α) and a high-gain activation function ($1/u_0$) for a given leak conductance $1/R$. On one hand, a high gain of the activation function is desired in order to minimize the activation cost and well approximate the original discrete WTA problem (3.3) using analog units. On the other hand, self-excitation has to be strong enough to give the units in the network enough gain to become potential winners. The self-excitation gain limit mutually constrains both effects. As a result, the total network gain might be insufficient to declare a clear winner, particularly for small input differences. Figure 3.4a shows the behavior of a WTA network with dynamics (3.12), simulated for three different inputs with α set at the self-excitation gain limit. For large input differences (bottom panel) the network is able to perform a *hard-WTA* competition resulting in a binary response pattern with the winning unit being active and all others inactive. However, if the input differences are smaller (middle and top panels) the gain is insufficient and the network only performs a *soft-WTA* competition: differences in the input are amplified but the gain is not strong enough to assign all activity to only one unit. An intuitive way of understanding this input-dependent behavior is that the network is not always able to fulfill the syntactic constraints. For small input differences, the γ term has only marginal influence and cannot significantly support the selection of the correct winner. The competition then remains mostly between the syntactic (α and β terms) and the activation constraint ($1/R$ term) which are conflicting. Because of the self-excitation gain limit, the syntactic constraints cannot clearly dominate the competition resulting in the observed network behavior. The soft-WTA behavior is a direct consequence of the unconstrained formulation of the WTA problem and is similar to the previously discussed scaling problem of the TSP.

Although the soft-WTA behavior does not always lead to exact solutions to the initially defined (hard-)WTA problem (3.2), it is a very useful computational property of the network. First of all, the network performance degrades gracefully under increasingly noisy conditions. This is important considering a physical implementation of the WTA network. If the signal-to-noise ratio (SNR) in the input differences becomes smaller, thus the reliability of the input decreases, the network increasingly "refuses" to select a winner. The output pattern of the network can be understood as expressing the confidence the network has in determining the correct winner. The more it approaches a binary output, the higher is its confidence in the selection of the winner. A hard-WTA operation would not be clever under low-SNR conditions. The network would always pick a winner but the selection would be dominated by the noise; a hard-WTA operation is not well-defined. Second, while a soft-WTA network promotes a binary selection of a winner, it still preserves part of the analog input. Thus, the output carries more information about the input distribution than just which input is largest. A soft-WTA network amplifies the relative differences in the input. It implements a differential amplifier circuit where the strength of self-excitation α determines the gain of the amplifier.

It has been widely argued that soft-WTA network architectures of the kind discussed here are a very plausible and simple model for many computational processes observed in biological neural networks. Multiplicative modulation of neural population responses such as gain fields (e.g. [Salinas and Abbott 1996]), response normalization, and complex cell properties (e.g. [Chance et al. 1999]) have been explained as a result of such competitive neuronal network structures. Also from a theoretical point of view it has been proven

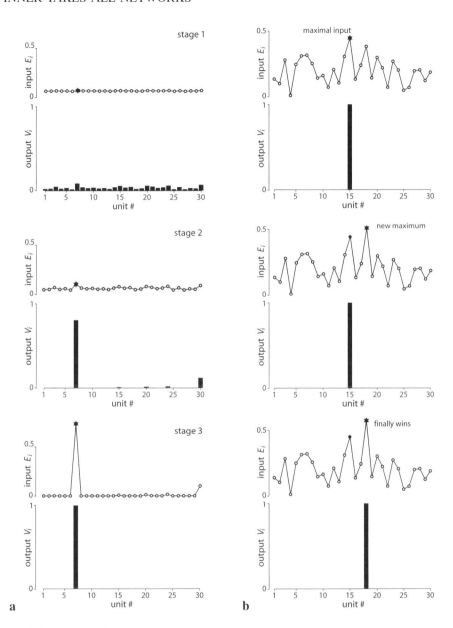

Figure 3.4 *WTA amplifier and hysteresis.*
(a) For small input differences (top and middle panel) the gain is too small to clearly select
a winner when obeying the self-excitation gain limit. Forming a cascade of WTA networks
by connecting the output of each stage with the input of the next stage (top to bottom)
multiplies the gain, leading to a hard-WTA solution without exceeding the self-excitation
gain limit in each stage. (b) Exceeding the gain limit leads to hysteretic behavior: a winner
is selected and kept although the input changes and a new unit receives the largest input
(middle panel). A substantially larger input is required to switch the network output to the
correct winner (bottom panel).

that a soft-WTA operation is a very powerful non-linear computational primitive; a single soft-WTA network has the same computational complexity as any complex multi-layer perceptron [Maass 2000].

Hysteresis

If α is set to exceed the self-excitation gain limit then the network has multiple local minima. Local minima typically express themselves as hysteresis, a multi-stable behavior of the network under dynamic input conditions. Figure 3.4b shows hysteresis in a WTA network where α is just slightly larger than the self-excitation gain limit. For the given input, the gain is large enough to well approximate a binary output with only the winning unit being active. Once the winning unit is assigned, however, it remains the winner even when the input changes and some other unit actually receives the largest input. A substantial input difference is required to overcome the local minimum and let the network switch to the correct winning input. This is not a correct solution of the original hard-WTA problem (3.2), yet such behavior can be of interest under some computational aspects. Hysteresis represents a simple form of memory that is somewhat resilient to noise. It keeps the last selection until a sufficiently different new input is presented. For example, the gain can be adjusted such that the significant difference exactly represents the expected SNR. Then, a new winner is only selected if it exceeds the old winner by the noise level.

Cascaded WTA amplifier

A cascade of WTA networks is an elegant solution to resolve the trade-off between hysteresis and limited gain. The idea is simple: the output of each WTA network along the cascade serves as the input to the next network. Such a cascaded WTA system can multiply the gain almost arbitrarily depending on the length of the cascade. It can do this with the self-excitation gain in each network being small enough to avoid hysteresis. Figure 3.4a, which was used previously to illustrate the soft-WTA behavior of the network, actually shows the behavior of a cascaded WTA with three stages. Each stage consists of an identical network (3.12) with 30 units. The input pattern to the first stage of the cascade (top panel) is almost uniform with input differences of $<1\%$. The star indicates the largest input. The gain is too small to clearly select a winner after the first stage. However, after passing the second stage the competition is decided and the output of the third stage, finally, well approximates a hard-WTA solution (bottom panel). The individual outputs of each stage are in synchrony, simultaneously providing different levels of amplification and winner selection for a given input.

3.4   Resistive Network

The example of the WTA network has demonstrated how a computational task can be formulated as an unconstrained optimization problem and how the appropriate network can be derived so that it solves the problem. The second example somewhat illustrates how to proceed in the opposite direction. Consider a given network of simple computational units, which resembles a well-known electrical structure. What is the computational problem it solves and how can it be formulated as an unconstrained optimization problem?

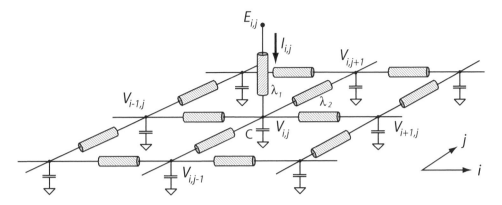

Figure 3.5 *Resistive network.*
For readability, only a single input node $E_{i,j}$ with input conductance λ_1 is shown.

The network to consider is shown in Figure 3.5. It is a homogeneous resistive (or dif-fusion) network on a square grid. The input to the network is represented by the voltage distribution $\vec{E}$ at the input nodes, whereas the output is considered to be the voltage distri-bution $\vec{V}$ at the nodes of the network. Each input node is connected to the output by some input conductance λ_1 and each network node is connected to its nearest neighbors by the lateral conductance λ_2. Note that Figure 3.5 only shows the input conductance of a single node. Applying Kirchhoff's current law leads to the following dynamics for a single node (i, j) in the network:

$$\dot{V}_{ij} = \frac{1}{C}[\lambda_1(E_{ij} - V_{ij}) + \lambda_2(V_{i,j+1} + V_{i,j-1} + V_{i+1,j} + V_{i-1,j} - 4V_{ij})]. \qquad (3.15)$$

The WTA example has shown that the dynamics of the network can be interpreted as performing gradient descent on a global cost function. Considering the right-hand side of (3.15) as the negative discrete partial derivative of such a cost function, it seems possible to reconstruct the cost function by partial integration. In general, there is no simple forward procedure to do this. In this simple case, however, an appropriate cost function can be found. The λ_2 term in (3.15) is the discrete five-point approximation of the Laplacian $\nabla^2 V_{ij}$ on a rectangular grid. It is straightforward to show that

$$H(V) = \underbrace{\frac{\lambda_1}{2} \sum_{ij}(E_{ij} - V_{ij})^2}_{\text{data term}} + \underbrace{\frac{\lambda_2}{2} \sum_{ij}((\Delta_i V_{ij})^2 + (\Delta_j V_{ij})^2)}_{\text{smoothing term}} \qquad (3.16)$$

is a valid cost function for the resistive network. $\Delta_{i,j}V_{ij}$ represents the discrete approxima-tion of the first-order spatial partial derivatives of V_{ij} in the i or j direction, respectively.

The cost function (3.16) is the combination of two constraints: the first constraint restricts the output V_{ij} not to be too far from the input E_{ij} (data term), while the sec-ond requires that the spatial variation of the output units be small (smoothness term). The relative weighting of both constraints is determined by the parameters λ_1, λ_2 that repre-sent the input and lateral conductances of the resistive network. For example, if the lateral conductances are zero ($\lambda_2 = 0$) each unit is disconnected from its neighbors. In steady state, the voltage on each node of the network perfectly matches its input voltage. Thus the cost (3.16) is minimal (zero) if $V_{ij} = E_{ij} \forall (i, j)$.

The resistive network is the simplest embodiment of *standard regularization*. Regularization techniques usually impose a data and a smoothness constraint on the solution. They have been widely proposed as a common framework to solve problems in early vision which usually require smooth spatial interaction [Poggio et al. 1985, Bertero et al. 1987, Yuille 1989, Szeliski 1996]. Note that in terms of its connectivity, the resistive network is a very efficient implementation of a convolution engine with simple exponential kernels.

4

Visual Motion Perception Networks

Chapter 2 proposed a basic outline for a visual motion perception system. I will now formulate the computational goal of that system as a constraint satisfaction problem, and then derive analog electronic network architectures that find the optimal solution which represents the estimated optical flow.

These networks are considered physical systems rather than methods or algorithms for the estimation of visual motion. A system is an entity built of individual parts that together implement a function. Typically, the function describes a dynamical relation between the input the system receives, and the output that it generates. Also, the function should be defined over the whole range of combinations of input, output, and internal states it can possibly encounter. Otherwise, a supervisor would be required to carefully monitor all states and detect those conditions for which the system does not work correctly; that is, those for which it is not defined. Under these conditions, the supervisor must bypass the system (possibly even reset it) and notify receivers of the output accordingly. It is obvious that such supervision is expensive to maintain and implement when dealing with the sparsely constrained real-world visual environment. Therefore, it is a basic requirement that these networks do not need a supervisor, hence that they are well-conditioned and do not need to be reset.

4.1 Model for Optical Flow Estimation

The *brightness constancy constraint* is the first constraint chosen. It defines the implicit matching model that relates changes in image brightness to visual motion. Because the brightness constancy constraint is defined in terms of the spatio-temporal brightness gradients, the input to the system is considered to be the gradients $E_x(x, y, t)$, $E_y(x, y, t)$, and $E_t(x, y, t)$ rather than the image brightness itself. It is evident that in an implementation of the system, the transduction of visual information and the extraction of the spatio-temporal gradients have to be carried out in a first processing step. For the time being and the theoretical analysis within this chapter, however, I will assume that the image brightness gradients are given. Phototransduction and gradient estimation will be discussed later, in the context of the actual silicon implementations. The system's output is considered to be the vector field $\vec{v}(x, y, t) = (u, v)$ representing the instantaneous estimate of the optical flow in image space.

The brightness constancy Equation (2.4) describes a line in velocity space representing the set of all possible velocity estimates with respect to the spatio-temporal brightness gradients. The brightness constancy constraint is defined as the function

$$F(\vec{v}(x, y, t)) = \left(E_x u + E_y v + E_t\right)^2 \tag{4.1}$$

and is a measure of how strongly a particular motion estimate deviates from the constraint line.[1] Since the Euclidean distance d between the constraint line and a point $P = (u_p, v_p)$ representing some motion estimate is given by

$$d(\vec{v}_p, E_x, E_y, E_t) = \frac{\left|E_x u_p + E_y v_p + E_t\right|}{\sqrt{E_x^2 + E_y^2}}, \tag{4.2}$$

the measure (4.1) can be seen as the square of this distance times the absolute brightness gradient. This implies that if the image contrast is high, and thus the signal-to-noise ratio (SNR) is high, the input is more reliable and a deviation of the motion estimate from the constraint line is less tolerated than when contrast is low. The brightness constancy constraint alone can be sufficient to define a least-squares error solution for global motion, given that the input is not uniform over the whole image.

The system's description is completed with a flow model that imposes assumptions about the properties of the expected optical flow. The first assumption is that the optical flow estimate is smooth. The *smoothness constraint*

$$S(\vec{v}'(x, y, t)) = u_x^2 + u_y^2 + v_x^2 + v_y^2 \tag{4.3}$$

defines a quadratic measure of how much the flow field varies across the image space. Smoothness assumes that locations in the image close to each other are more likely to belong to the same moving object, and thus are more likely to have the same visual motion. It can been shown that smoothness is immediately related to the motion field of rigid objects [Ullman and Yuille 1987]. Conditions where the smoothness constraint is clearly violated are found at object boundaries, locations where motion discontinuities are likely to occur. Motion discontinuities will be discussed later. The smoothness constraint (4.3) has been proposed in previous models of visual motion estimation [Horn and Schunck 1981, Hildreth 1983], and is widely applied in regularization approaches of other early vision problems [Poggio and Torre 1984, Poggio et al. 1985]. Equation (4.3) requires the

---

[1]The brightness constancy constraint has an interesting relation to the flow dynamics in fluids. If the brightness distribution is identified with the density distribution of an incompressible fluid, then the brightness constancy Equation (2.4) describes the two-dimensional dynamics of the fluid under the assumption of preserved total mass. This analogy suggests that the brightness constancy constraint holds also for non-rigid objects as long as the total brightness (total mass) of the object stays constant. The restriction $\text{div}(\vec{v}) = 0$ follows directly from the assumption of constant mass.

optical flow field to be differentiable. For convenience, it is required to be at least twice continuously differentiable. The smoothness constraint has a physical equivalent which is the membrane model. The minimal integral of (4.3) over an area $\Omega \in \mathbb{R}^2$ corresponds to the minimal kinetic energy of a membrane. A string model, which is a one-dimensional membrane, will later illustrate the complete optimization problem (Figure 4.1). Smoothness could be enforced by other constraints (*e.g.* averaging the brightness constancy constraint over a small image area), but applying the above formulation permits a very compact implementation using resistive networks.

Since (4.3) constrains the derivatives and not the optical flow field itself, there remain input conditions for which the brightness constancy and the smoothness constraint do not sufficiently constrain the estimation problem. For example, if the image brightness is uniform then the brightness gradients are all zero, and the brightness constancy constraint is trivially fulfilled. Consequently, the smoothness constraint does not define a unique optimal solution because there are an infinite number of spatially uniform flow fields that are equally (and maximally) smooth. The optimization problem is ill-posed. By forcing an additional constraint on the solution expressed as the *bias constraint*

$$B(\vec{v}(x, y, t)) = (u - u_{\text{ref}})^2 + (v - v_{\text{ref}})^2, \tag{4.4}$$

one can guarantee a well-posed optimization problem. This is the second assumption of the flow model. The bias constraint measures how much the estimated optical flow vector deviates from a given reference motion (u_{ref}, v_{ref}). Besides forcing the system always to have a unique solution, the bias constraint has a more fundamental meaning. It permits the system to incorporate and apply a priori information about the expected motion. The reference motion vector $\vec{v}_{\text{ref}} = (u_{\text{ref}}, v_{\text{ref}})$ specifies ideally a good optical flow estimate under conditions where the visual input is ambiguous and unreliable. It is a parameter of the optimization problem that can be used to adapt the system to different visual conditions and needs. For example, the value of the reference motion vector could be adapted to the statistical properties of the visual world, for example the statistical mean of visual motion over time. Or, the system sees motion of a particular class of objects of which it knows its tendency to move with a particular motion. Thus, the bias constraint incorporates such additional knowledge, leading to a better estimate of visual motion.

Finally, the three constraints above are combined to express the following unconstrained optimization problem:

Given the continuous input $E_x(x, y, t)$, $E_y(x, y, t)$, and $E_t(x, y, t)$ on an image region $\Omega \subset \mathbb{R}^2$, find the optical flow field $\vec{v}(x, y, t) \in \mathbb{R}^2$ such that the cost function

$$H(\vec{v}(x, y, t); \rho, \sigma) = \int_{\Omega} (F + \rho S + \sigma B) \, d\Omega \tag{4.5}$$

is minimal, with

$$\vec{v}'(x, y, t) = \vec{0} \quad \text{along } \partial\Omega ; \quad \text{and } \sigma > 0, \text{ and } \rho \geq 0. \tag{4.6}$$

The relative weights of the constraints are set by the parameters ρ and σ. In the first instance, these parameters are assumed to be constant. However, they will be considered later to be a function of time, space, and the current optical flow estimate, and so permit the computational behavior to be altered significantly. The weighting parameters and the reference motion provide the possibility to program the system in a limited way to account for different models of visual motion perception.

The optical flow gradients are assumed to vanish at the boundary $\partial\Omega$; that is, that the area outside Ω does not influence the motion estimate. This assumption seems reasonable, because the system does not have any information available outside the image space.

4.1.1   Well-posed optimization problem

Before considering an appropriate network solution, it is important to ensure that the constraint satisfaction problem always has a solution which is unique, and depends continuously on the input; thus, that it is *well-posed* according to the definition of Hadamard. The following theorem proves this (more details can be found in Appendix A).

Theorem.   *The estimation of optical flow formulated as the constraint optimization problem (4.5) is well-posed. In particular, this holds under all possible visual input conditions.*

Proof.   Consider the integrand

$$L(\vec{v}(x, y, t), \vec{v}(x, y, t)', x, y; \rho, \sigma) = F + \rho S + \sigma B \tag{4.7}$$

of the cost function (4.5). L is continuously defined for all $\vec{v}(x, y, t) \in \mathbb{R}^2$ that are sufficiently smooth. Hence, each twice continuously differentiable flow field that fulfills the boundary condition is a candidate for a solution. According to proposition 1 in Appendix A, the variational problem (4.5) has a unique solution $\vec{v}_0$ if the associated Euler–Lagrange equations have a solution, the integrand L is strictly convex, and the natural boundary condition

$$L_{\vec{v}'}(\vec{v}_0(x, y, t), \vec{v}_0(x, y, t)', x, y) = 0 \tag{4.8}$$

holds. For (4.5), the associated Euler–Lagrange equations are

$$\begin{aligned}
\rho(u_{xx} + u_{yy}) &= E_x(E_x u + E_y v + E_t) + \sigma(u - u_{\text{ref}}) \\
\rho(v_{xx} + v_{yy}) &= E_y(E_x u + E_y v + E_t) + \sigma(v - v_{\text{ref}}).
\end{aligned} \tag{4.9}$$

They represent a typical system of stationary, inhomogeneous Poisson equations. For the required boundary conditions (4.6), the existence of a solution is guaranteed [Bronstein and Smendjajew 1996].

Strict convexity of the integrand is given if the Hessian $\mathsf{J} = \nabla^2 L(\vec{v}, \vec{v}', x, y)$ is positive definite, thus all eigenvalues are real and positive. The Hessian of the integrand L is

$$\mathsf{J} = \begin{pmatrix}
E_x^2 + \sigma & E_x E_y & 0 & 0 & 0 & 0 \\
E_y E_x & E_y^2 + \sigma & 0 & 0 & 0 & 0 \\
0 & 0 & \rho & 0 & 0 & 0 \\
0 & 0 & 0 & \rho & 0 & 0 \\
0 & 0 & 0 & 0 & \rho & 0 \\
0 & 0 & 0 & 0 & 0 & \rho
\end{pmatrix}. \tag{4.10}$$

The matrix J is symmetric and has three real and distinct eigenvalues

$$\lambda_1 = \rho, \quad \lambda_2 = \sigma, \quad \text{and} \quad \lambda_3 = \sigma + E_x^2 + E_y^2. \tag{4.11}$$

Since the weighting parameters ρ and σ are positive, the eigenvalues are always positive independently of the visual input. Therefore, the integrand L is strictly convex. It is straightforward to show that the optical flow gradient vanishes at the image boundaries due to the natural boundary condition (4.8) since

$$L_{\vec{v}'} = 2\rho\, \vec{v}(x, y, t)' \equiv \vec{0} \qquad \text{only for} \quad \vec{v}'(x, y, t) = \vec{0}. \tag{4.12}$$

Thus all conditions of proposition 1 are fulfilled and the optimization problem has a global minimum and therefore a unique solution. Again, this is truly independent of the brightness gradients and thus the visual input.

Finally, continuity of the solution on the input is guaranteed because L is a continuous function of $\vec{v}(x, y, t)$ and the continuous spatio-temporal gradients $E_x(x, y, t)$, $E_y(x, y, t)$, and $E_t(x, y, t)$. Thus the proof is complete. □

Boundary conditions

If σ is set to zero and therefore the bias constraint loses its influence and is abandoned, the cost function (4.5) is equivalent to the optical flow formulation of Horn and Schunck [1981]. Now, at least one eigenvalue (4.11) of the Hessian is zero. The cost function is still convex but no longer strictly convex and thus multiple minima might co-exist. Local minima occur, for example, when the input is spatially uniform. In this case the problem is ill-posed. Not surprisingly some authors [Poggio et al. 1985] suggest restricting the visual input space in order to guarantee well-posedness. Such restrictions, however, require a supervisor which is, as mentioned before, considered not to be a valuable option. Well-posedness can be restored by defining rigorous boundary conditions [Horn 1988]. For example, Koch et al. [1991] proposed to set the flow estimate $\vec{v}(x, y, t)$ to zero along the image boundary $\partial\Omega$. Clearly, a fixed motion value at the image boundary cannot be justified by any reasonable assumption other than to prevent ill-posedness. It greatly influences the estimation of optical flow near the image boundary and implies an unnecessarily strong bias.

The bias constraint (4.4), however, ensures well-posedness while allowing natural boundary conditions. Formulating an unconstrained optimization problem also permits a trade-off between the satisfaction of the individual constraints. Thus, depending on the value of parameter σ, it permits us to express a more or less strong tendency rather than a commitment to some reference motion $\vec{v}_{\text{ref}}$.

4.1.2 Mechanical equivalent

Figure 4.1 represents a mechanical model equivalent of the optimization problem (4.5). It illustrates in an intuitive way the influence of the different constraints and how they interact. In this model equivalent, the position of a string above some ground level represents the estimated optical flow. To improve clarity of the illustration, the problem is considered only in one spatial dimension. Pins represent the local optical flow estimate for which the measured brightness constancy constraint is fulfilled. Note that in the one-dimensional case

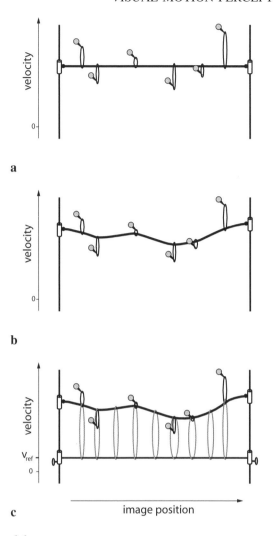

Figure 4.1 *String model.*
The string model is a one-dimensional mechanical model equivalent to illustrating the influence of the different constraints. The optical flow estimate is represented by the position of the string (one-dimensional membrane) relative to some resting level. Pins indicate the positions where the local brightness constancy constraints are fulfilled. (a) Rubber bands enforce the local brightness constancy constraints. The elasticity constant of each individual rubber band is proportional to the local brightness gradient. If the string is completely stiff, the result is a total least-squares error solution for global flow. (b) Adding the smoothness constraint is equivalent to allowing the string to bend. The tension of the string is given by the weighting parameter ρ. Obviously, for no visual input, the rubber band strengths are zero and the string floats, which is an ill-posed condition. (c) Introducing the bias constraint is equivalent to introducing another set of (weak) rubber bands connecting the string to the reference level V_{ref}. The elastic constant of these bands is proportional to the parameter σ. Now, the position and shape of the string are always well-defined.

this is a point rather than a line. Rubber bands enforce the brightness constancy constraint by pulling the string toward the pins. The elasticity constant of each band is proportional to the local gradient of the brightness constancy constraint. The boundary conditions (4.6) are imposed by allowing the ends of the string to move freely along the vertical axis. The first case (Figure 4.1a) illustrates the least-squares error solution for a global flow estimate. The tension of the string ρ is infinitely high and so it is completely stiff; its position represents the global estimate of the optical flow. Introducing the smoothness term, so assuming a finite ρ, allows the string to bend and to provide a smooth, local optical flow estimate. In both cases (Figures 4.1a and b), however, the string floats freely and the problem becomes ill-posed if no visual input is available and thus no pins with rubber bands constrain the string's position. On adding the bias constraint (Figure 4.1c) this is no longer possible. The bias constraint is applied with the aid of another set of rubber bands with elasticity constants given by the parameter σ. This second set pulls the string toward the reference level v_{ref}. Even when there is no visual input (no pins), it will constrain the position of the string, forcing it to be at the reference level.

The analogy between the above string model and the optimization problem (4.5) is not perfect. In the string model, the brightness constancy and the bias constraint are only imposed at sparse spatial locations. This has some implications that are discussed in the next section. Nevertheless, the string model illustrates well the function and effect of the individual constraints.

4.1.3 Smoothness and sparse data

The so-called *motion coherence theory* [Yuille and Grzywacz 1988, Yuille and Grzywacz 1989, Grzywacz and Yuille 1991] proposes a general formulation of the optical flow estimation problem that reveals some similarities with the optimization problem (4.5). Yuille and Grzywacz proposed the minimization of the following cost function in order to obtain an optimal estimate of the optical flow:

$$H_{GY}(\vec{v}(\vec{r}), \vec{U}_i) = \sum_i (M(\vec{v}(\vec{r}_i)) - M(\vec{U}_i))^2 + \lambda \int d\vec{r} \sum_{m=0}^{\infty} c_m (D^m \vec{v})^2. \qquad (4.13)$$

The first term of the cost function represents the difference between the local measurement of visual motion $M(\vec{U}_i)$, where U_i is the *true*[2] image motion as defined by the applied matching model, and the local value of the estimated smooth flow field $M(\vec{v}(\vec{r}_i))$, summed over all locations r_i where visual input is available. Yuille and Grzywacz assume that the measurements are sparse and only available at some spatial locations i, whereas the estimated optical flow $\vec{v}(\vec{r})$ is defined on continuous space. The first term represents a generic way of expressing a matching constraint. The second term is a description of a general smoothness constraint that requires the derivatives ($D^m \vec{v}$, $m \in \mathbb{Z}^+$) of the optical flow field of any order m to be small. The zero-order derivatives are defining a particular form of the bias constraint (4.4) assuming the reference motion (u_{ref}, v_{ref}) to be zero. The c_m's are functions that can be chosen but should increase monotonically.

---

[2]Original formulation [Grzywacz and Yuille 1991]; the notion of *true* image motion is unfortunate because, as discussed, there is no true image motion. True refers here to the motion estimate that is in agreement with the applied matching model.

It is important to note that the cost function (4.13) assumes sparse data, that is visual information is only given at discrete locations $\vec{r}_i$. This has the effect that a smoothing term must be included in order to make the problem well-posed [Duchon 1977]. According to Duchon, regularization over sparse data in n-dimensional space $\mathbb{R}^n$ requires derivatives of order m, where $m > n/2$. The previous, one-dimensional string model ($m > 1/2$) can give some intuition why this is case. Theoretically, zero-order derivatives are not enough to constrain the solution. Without the smoothness constraint (first-order derivative) the string has no tension and thus it follows the reference motion v_{ref} everywhere except at the sparse data points where it is a δ-function. Thus, a continuous estimate $\vec{v}(\vec{r})$ is not defined. In a physical world, however, a string is never ideal. It always has finite thickness and thus, in the limit ($\rho = 0$), acts like a rod constraining smoothness via higher order derivatives. It naturally remains a well-defined system.

Higher than zero-order derivatives, however, are *not* required when the brightness distribution $E(x, y, t)$ is continuously differentiable. Also, they are not necessary if image brightness and optical flow are both defined only on a discrete (sparse) spatial lattice. The analysis for the continuous space applies to the discrete subspace without restrictions as long as the derivatives are properly defined. In fact, visual processing problems are almost always defined on a spatial lattice because they usually begin with images which are discrete samples of the brightness distribution. The bias constraint, on the other hand, is always necessary if the measurements of the matching model are ambiguous or undefined. This is independent of the sparseness of the visual data. In the optical flow model (4.5), the measurements are expressed by the brightness constancy constraint which is ambiguous with respect to the optical flow estimate. Thus any model that relies on the brightness constancy constraint needs a bias constraint of the form (4.4); for example, the Horn and Schunck [1981] algorithm is ill-defined not because it does not apply smoothing in second-order derivatives, but because it does not apply a bias constraint.

The motion coherence theory defines a rather special optimization problem and is not a practical algorithm for visual motion estimation. It was intended to describe the process of motion capture and motion integration with a focus on random dot stimuli. In particular, it is based on the assumption that the system is always able to perform an unambiguous local measurement $M(\vec{U}_i)$ of visual motion. The limitations and restrictions of this assumption have been discussed before. Thus, care has to be taken when generalizing some of the theorems of motion coherence theory to other formulations of visual motion estimation. As shown, some of them have their origin in the particular assumptions and definitions of this theory.

4.1.4 Probabilistic formulation

Constraint optimization (regularization) is one way to address the ambiguous and noisy problem of estimating visual motion. A probabilistic interpretation is another. For example, the optimization problem (4.5) can be expressed as the conditional probability distribution over all possible optical flow fields given the input

$$P[\vec{v}|E_x, E_y, E_t] \propto \exp(-H(\vec{v}; \sigma, \rho)), \tag{4.14}$$

where $H(\vec{v}; \sigma, \rho)$ is the cost function (4.5). The solution that minimizes the cost is the one with highest probability. The probability formulation becomes more meaningful when considering $P[\vec{v}|E_x, E_y, E_t]$ as the *posterior* of a Bayesian estimator. Bayes' theorem

$$P[\vec{v}|E_x, E_y, E_t] \;\propto\; P[E_x, E_y, E_t|\vec{v}]P[\vec{v}] \tag{4.15}$$

states that the posterior, thus the probability that the observed spatio-temporal brightness gradient is generated by a particular optical flow pattern, is proportional to the *likelihood* $P[E_x, E_y, E_t|\vec{v}]$, the probability of a particular flow field generating the observed visual input,[3] times the *prior* $P[\vec{v}]$, which is the probability of such a particular optical flow field occurring. The intuitive understanding of such a Bayesian estimator is the following: the likelihood represents the quality of the visual information, thus how well the measured spatio-temporal brightness gradients constrain a particular optical flow estimate. The lower the quality of the gradient measures and the less they constrain the estimate, the broader the likelihood function, and the more the estimator relies on stored a priori information represented by the prior. Picking the optical flow estimate which maximizes the posterior probability is then often an optimal strategy. A Bayesian interpretation is a sensible way to describe early vision estimation problems because they often have to deal with ambiguous and noisy visual input [Szeliski 1990]. A number of authors have described a Bayesian estimation framework for the problem of estimation optical flow [Konrad and Dubois 1992, Simoncelli 1993, Weiss and Fleet 2002].

For the optimization problem (4.5), the equivalent likelihood is defined by the brightness constancy constraint. At each location (x_i, y_i), the quadratic norm of the constraint leads to a local likelihood function that is Gaussian in $d(\vec{v}_i, E_{x_i}, E_{y_i}, E_{t_i})$, the distance between the optical flow $\vec{v}_i$ and the local brightness constancy constraint line. The width of the Gaussian σ_l is inversely proportional to the total local brightness gradient $(E_{x_i}^2 + E_{y_i}^2)$ because the brightness constancy constraint defines the square of the distance (4.2) normalized by the total brightness gradient. Note that normalization emerges directly from the brightness constancy constraint. The combined local brightness gradient can be interpreted as the local image contrast. The likelihood function over the entire image space is then given as

$$P[E_x, E_y, E_t|\vec{v}] \propto \exp\left(-\int_{x,y} \frac{d^2(\vec{v}, E_x, E_y, E_t)}{2\sigma_l^2}\right). \tag{4.16}$$

The prior distribution of the Bayesian estimator is defined by the smoothness and the bias constraint. The prior reflects the expectations the estimator has for the estimated flow field. In this case, the optical flow field is expected to be smooth and close to some reference motion. The smoothness assumption is based on the rigidity of the observed objects, whereas the particular form of the bias constraint so far has been motivated only in the context of well-posedness. Yet, it turns out that smoothness and a bias toward slow motions are explaining well the human perception of visual motion[4] [Weiss and Adelson 1998, Weiss et al. 2002]. The prior can be written as

$$P[\vec{v}] \;\propto\; \exp[-(S + B)] \tag{4.17}$$

with S and B given by (4.3) and (4.4).

---

[3]Note that Bayes' rule requires us to think of the likelihood as a function of the conditioning variable $\vec{v}$. The likelihood is not a probability density over the flow field.

[4]In a recent study, we have shown that a Gaussian prior and thus a quadratic norm of the bias constraint is quantitatively not exactly the right choice to explain human perception. It seems rather that humans have a prior that approximately follows a power-law distribution [Stocker and Simoncelli 2005a]

The particular form of the smoothness constraint (4.3) does not permit us to formulate an independent Bayesian estimator of the optical flow for each image location, which would substantially simplify the computational problem, unless $\rho = 0$. Some alternative way to impose smoothness is to discard the smoothness constraint (4.3) as is, but instead to integrate the brightness constancy constraint over small and smooth isotropic spatial kernels ω [Lucas and Kanade 1981, Weiss et al. 2002]. In this way the posterior becomes only a distribution of the local optical flow vector $\vec{v}_i$, thus

$$P[\vec{v}_i | E_x, E_y, E_t] \propto \exp\left(-\int_{\omega} (E_x u_i + E_y v_i + E_t)^2 - \sigma((u_i - u_{\text{ref}})^2 + (v_i - v_{\text{ref}})^2)\right).$$
(4.18)

For each location (x_i, v_i), the most probable estimate, hence the *maximum a posteriori (MAP)* solution, can be written in closed form.

The Bayesian formulation and the constraint optimization problem are equivalent. The optimal solutions are identical; the global minimum of the optimization problem is equivalent to the MAP, and convexity relates to log-concavity. It has been argued that it might be advantageous for a visual motion system not only to compute a single solution (e.g. MAP) but to explicitly store the entire probability distributions. It is interesting to think about how one would represent and compute such distributions in networks such as the brain [Rao 2004, Pouget et al. 2003, Hoyer and Hyvarinen 2002]. However, the benefit is not obvious, in particular when the task only requires a deterministic answer as is the case in the optical flow estimation problem; the optimal solution can always be described in a deterministic way.

The Bayesian formulation, however, provides a different understanding of the model constraints in terms of meaningful probabilistic components. Specifically, its likelihood function expresses the relationship between the measured spatio-temporal brightness gradients and the flow field to be estimated, and reflects measurement noise and visual ambiguities induced by the visual stimulus. The prior distribution expresses the expectation of the observer to encounter a particular image motion. The expectations of an optimal observer should match the statistics of the image motion generated by its environment.

4.2 Network Architecture

So far, the optimization problem (4.5) has been formulated on the continuous image space $\Omega \subset \mathbb{R}^2$. However, because the visual input is usually provided by some imaging device with discrete spatial resolution, it is feasible to integrate the cost function only over locations where the input is accessible. Consider a discrete image space defined on an orthogonal lattice, where each node is labeled by two integer values $i \in [1 \ldots n]$ and $j \in [1 \ldots m]$. On this discrete space, the estimation of optical flow is reduced to a finite number of locations. Yet, the previous analysis of the optimization problem does apply to the discrete case without restrictions as long as the optical flow gradients are defined properly. On a discrete lattice, the cost function (4.5) changes to

$$H(\vec{v}; \rho, \sigma) = \sum_{i=1}^{n} \sum_{j=1}^{m} \left[(E_{xij} u_{ij} + E_{yij} v_{ij} + E_{tij})^2 \right.$$
$$\left. + \rho((\Delta u_{ij})^2 + (\Delta v_{ij})^2) + \sigma((u_{ij} - u_{\text{ref}})^2 + (v_{ij} - v_{\text{ref}})^2) \right]$$
(4.19)

where the partial derivatives of the flow vectors in the smoothness term are replaced by the difference operator[5] Δ such that

$$(\Delta x_{i,j})^2 = \left(\frac{x_{i+1,j} - x_{i,j}}{2h}\right)^2 + \left(\frac{x_{i,j+1} - x_{i,j}}{2h}\right)^2.$$

The optimization problem has a unique solution as in the continuous case. Therefore, it follows directly that

$$H'(\vec{v}; \rho, \sigma) = 0 \qquad (4.20)$$

is a sufficient condition for the unique solution. Similarly to the WTA example discussed previously, a dynamical network is defined that performs *gradient descent* on the cost function (4.19). The network changes the estimated optical flow components $u_{i,j}$ and $v_{i,j}$ at each node negatively proportional to the partial gradients of the cost function thus,

$$\dot{u}_{ij} \propto -\frac{\partial H(\vec{v}; \rho, \sigma)}{\partial u_{ij}} \quad \text{and} \quad \dot{v}_{ij} \propto -\frac{\partial H(\vec{v}; \rho, \sigma)}{\partial v_{ij}} \qquad (4.21)$$

until steady state is reached. This leads to the following system of $2n \times m$ linear partial differential equations:

$$
\begin{aligned}
\dot{u}_{ij} &= -\frac{1}{C}\Big[E_{xij}(E_{xij}u_{ij} + E_{yij}v_{ij} + E_{tij}) \\
&\quad - \rho(u_{i+1,j} + u_{i-1,j} + u_{i,j+1} + u_{i,j-1} - 4u_{ij}) + \sigma(u_{ij} - u_{\text{ref}})\Big], \\
\dot{v}_{ij} &= -\frac{1}{C}\Big[E_{yij}(E_{xij}u_{ij} + E_{yij}v_{ij} + E_{tij}) \\
&\quad - \rho(v_{i+1,j} + v_{i-1,j} + v_{i,j+1} + v_{i,j-1} - 4v_{ij}) + \sigma(v_{ij} - v_{\text{ref}})\Big]
\end{aligned}
\qquad (4.22)
$$

for all $i \in [1 \ldots n]$, $j \in [1 \ldots m]$ and where C is a constant. These dynamics (4.22) define the optical flow network.

Again, Equations (4.22) are interpreted as describing the voltage dynamics in an electronic network with two cross-coupled resistive layers. Figure 4.2 sketches a single unit of such an electronic network. For each unit, the two estimated components of the optical flow vector are represented by the voltages U_{ij} and V_{ij} on a capacitance C with respect to some virtual ground V_0. Equations (4.22) formulate Kirchhoff's current law: the terms in square brackets represent the total currents that flow onto each of the two nodes, and which balance the capacitive currents $C\dot{u}$ and $C\dot{v}$, respectively. At steady state, both the capacitive currents and currents flowing onto the nodes become zero. If this is the case for every node in the network then the voltage distributions represent the optimal optical flow estimate according to the cost function (4.19).

Each constraint of the cost function is physically instantiated in the following way:

- Smoothness is enforced by the resistive layers with lateral conductances ρ.

---

[5]Not to be confused with the Laplace operator ∇^2.

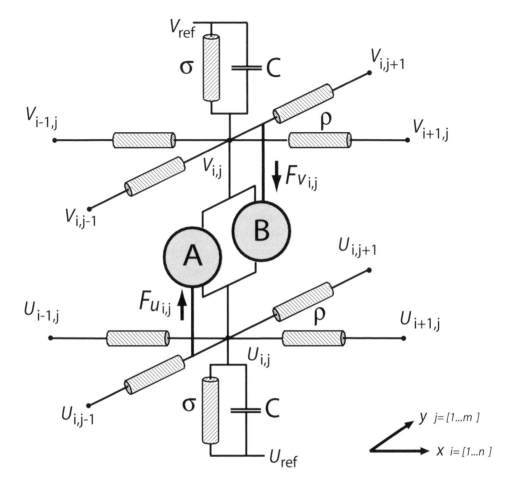

Figure 4.2 *Optical flow network.*
A single unit of the optical flow network is shown. The smoothness constraint is embodied
in two resistive layers composed of conductances ρ. The components of the local optical
flow vector are encoded as the voltages $U_{i,j}$ and $V_{i,j}$ at each node of these layers, relative
to some virtual ground V_0. The vertical conductance σ accounts for the bias of the optical
flow estimate toward some reference motion represented by the potentials U_{ref} and V_{ref}.
A and B constitute active circuitry that acts as a current source or sink depending on the
deviation of the optical flow estimate from the local brightness constancy constraint.

- The bias constraint is implemented as a leak conductance σ to some reference potentials V_{ref} and U_{ref}.

- The realization of the brightness constancy constraint requires some active circuitry
  represented by the "constraint boxes"[6] A and B. At each node they sink currents

---

[6]A term coined by Harris and colleagues [Harris et al. 1990b].

Fu_{ij} and Fv_{ij}, which are proportional to the gradient of the brightness constancy constraint, thus

$$Fu_{ij} \propto \partial F_{ij}/\partial u_{ij} = -E_{xij}(E_{xij}u_{ij} + E_{yij}v_{ij} + E_{tij})$$
$$Fv_{ij} \propto \partial F_{ij}/\partial v_{ij} = -E_{yij}(E_{xij}u_{ij} + E_{yij}v_{ij} + E_{tij}). \tag{4.23}$$

These currents instantiate negative feedback because they are negatively proportional to the voltages on the capacitive node. The circuitry within these boxes includes several multiplication and summing operations. Since the components of the optical flow vector can take on negative values, the voltage encoding is relative to some virtual ground V_0.

The total dissipated power in the two resistive layers is the physical equivalent of the total cost induced by the bias and smoothness constraint. For constant conductances ρ and σ the behavior of the resistive network is ohmic, and the dissipated power is equivalent to the current flowing through the conductances times the potential difference across a node.

4.2.1   Non-stationary optimization

Under the assumption of stationary input, the gradient descent dynamics continuously decrease the cost function (4.19) and the system globally converges to a single solution as $t \to \infty$. $H(\vec{v}; \rho, \sigma)$ is said to be a Lyapunov function of the system, that is

$$\frac{\mathrm{d}H}{\mathrm{d}t} \leq 0 \qquad \text{with} \qquad \frac{\mathrm{d}H}{\mathrm{d}t} = 0 \quad \text{only for} \quad \dot{\vec{v}} = 0. \tag{4.24}$$

This is verified by expanding the total derivative of the cost function and substituting in the dynamics (4.22), hence

$$\frac{\mathrm{d}H}{\mathrm{d}t} = \sum_{ij} \left[\frac{\partial H}{\partial u_{ij}} \dot{u}_{ij} + \frac{\partial H}{\partial v_{ij}} \dot{v}_{ij} \right]$$

$$= -\frac{1}{C} \sum_{ij} \left[\left(\frac{\partial H}{\partial u_{ij}} \right)^2 + \left(\frac{\partial H}{\partial v_{ij}} \right)^2 \right] \leq 0. \tag{4.25}$$

However, the visual input is hardly stationary, of course, because the estimation of optical flow is per se defined on the basis of spatio-temporal changes in image brightness. The estimation problem becomes non-stationary and it is important to investigate the consequences for the gradient descent strategy of the optical flow network. In the non-stationary case, the total temporal derivative of the cost function becomes

$$\frac{\mathrm{d}H}{\mathrm{d}t} = \sum_{ij} \left[\frac{\partial H}{\partial u_{ij}} \dot{u}_{ij} + \frac{\partial H}{\partial v_{ij}} \dot{v}_{ij} + \frac{\partial H}{\partial E_{xij}} \frac{\mathrm{d}E_{xij}}{\mathrm{d}t} + \frac{\partial H}{\partial E_{yij}} \frac{\mathrm{d}E_{yij}}{\mathrm{d}t} + \frac{\partial H}{\partial E_{tij}} \frac{\mathrm{d}E_{tij}}{\mathrm{d}t} \right]. \tag{4.26}$$

Although the first two terms are less than or equal to zero as shown above, this is not necessarily the case for the total expression. The cost function is no longer a Lyapunov function by its strict definition. However, if the dynamics of the network are much faster than the input dynamics (thus *e.g.* $\dot{u}_{ij} \gg \mathrm{d}E_{xij}/\mathrm{d}t$), then the last three terms in (4.26) become negligible. Hence, the optimization problem can be considered quasi-stationary. In this case it is safe to assume that the solution is always at the momentary global minimum of the cost function – or at least very close to it. Consider again the mechanical analogy of the string shown in Figure 4.1, and assume now that the position of the pins, and thus the external forces applied by the rubber bands, change over time. Then, the shape of the string immediately adapts to the new configuration so that it will always represent the state of minimal costs. A critical limit is reached when the changes in pin positions happen on such a short time-scale that the mass of the string becomes relevant for its dynamics, and thus the kinetic energy of the system must be considered. The dynamics in the optical flow network are mainly determined the time constant τ_{rec} of the active recurrent circuitry represented by the boxes A and B in Figure 4.2. The time constant is to some extent under the active control of the designer of the system, and therefore can be minimized. The dynamics will be addressed again later when discussing processing speeds of the actual aVLSI implementation of the network (see Section 6.5).

4.2.2   Network conductances

The computational characteristics of the network at steady state mainly depend on the relative impact of the three different constraints and thus on the values of ρ and σ. According to these values, the computational behavior of the system changes significantly and exhibits different models of visual motion estimation such as normal flow, smooth local flow, or global flow. Figure 4.3 schematically illustrates the different computational behavior of the system as a function of the two parameters. The shaded oval region represents the parameter space in which the system provides robust and sensible optical flow estimation. The limits of sensible operation are determined by the bias conductance σ. For values approaching zero, computation can become ill-posed, whereas for high values, the behavior is trivial because the bias constraint dominates such that the system constantly reports the reference values.

The notion of computational complexity of the system is interesting. The contour plot in Figure 4.3 qualitatively illustrates the different levels of computational complexity. Here, complexity is meant to approximately indicate the computational load (execution time) needed if the solutions of the optical flow network were simulated on a sequential digital computer. Normal or global flow is computationally least expensive because these estimates can be described by mathematically closed-form expressions. Using sequential computers such feedforward computation needs much less time than the highly iterative process required for smooth optical flow estimation which is therefore considered computationally most complex. However, it is not obvious how to define an equivalent measure of the computational load for the optical flow network. Clearly, the active processing in each unit of the network does not change for different values of σ and ρ. So the processing load for each unit remains the same independently of whether it is estimating normal flow or contributing to a global flow estimate. The only difference in behavior is induced by the parameter ρ. Thus, what an observer of the system finally recognizes as a computational behavior of different complexity reduces to the amount of interaction within the network!

In the following, the different computational behaviors in relation to different values of the network conductances ρ and σ are considered in more detail. An artificial image

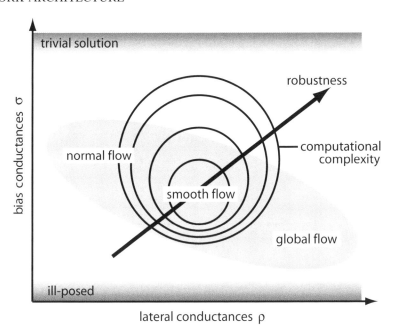

Figure 4.3 *Computational behavior depending on ρ and σ.*
The computational characteristics of the optical flow network are schematically illustrated as a function of the network conductances ρ and σ. None or very weak lateral coupling combined with a finite value of σ leads to normal flow estimation. On the other hand, a larger ρ results in a global estimate of optical flow. Robustness typically increases with larger integration areas (large ρ). Larger values of σ also increase robustness but are not desired because they strengthen the estimation bias significantly. The system shows its computationally richest and most complex behavior for small σ's and intermediate ρ's. This is indicated by the symbolic contour plot showing the different levels of computational complexity.

sequence is used for illustration, consisting of a triangle-shaped object moving on a stationary either lightly structured or unstructured background. A single frame of each sequence is shown in Figure 4.5a and b below.

Smoothness

The conductance ρ directly controls the smoothness of the optical flow estimate. If $\rho = 0$, the lateral coupling between the units in the network is disabled. There is no communication between individual units and no spatial integration of visual information takes place. The network performs a strict local optical flow estimate where each unit reports the flow vector that is closest to its brightness constancy constraint line and best matches the reference motion $\vec{v}_{\mathrm{ref}}$ enforced by the bias constraint. By suppressing lateral interactions, the network breaks up into an array of independent local processors. For any non-zero σ, the steady-state output of each unit can be computed in closed form by applying the necessary condition

$H'(\vec{v}; \rho, \sigma) = 0$, thus

$$E_{xij}(E_{xij}u_{ij} + E_{yij}v_{ij} + E_{tij}) + \sigma(u_{ij} - u_{\text{ref}}) = 0$$

$$E_{yij}(E_{xij}u_{ij} + E_{yij}v_{ij} + E_{tij}) + \sigma(v_{ij} - v_{\text{ref}}) = 0. \tag{4.27}$$

Solving for the local flow vector $\vec{v}_{ij}$ leads to

$$u_{ij} = -\underbrace{\frac{E_{tij}E_{xij}}{\sigma + E_{xij}^2 + E_{yij}^2}}_{\text{normal flow}} + \underbrace{\frac{u_{\text{ref}}(\sigma + E_{yij}^2) - v_{\text{ref}}(E_{xij}E_{yij})}{\sigma + E_{xij}^2 + E_{yij}^2}}_{\gamma}$$

$$v_{ij} = -\underbrace{\frac{E_{tij}E_{yij}}{\sigma + E_{xij}^2 + E_{yij}^2}}_{\text{normal flow}} + \underbrace{\frac{v_{\text{ref}}(\sigma + E_{xij}^2) - u_{\text{ref}}(E_{xij}E_{yij})}{\sigma + E_{xij}^2 + E_{yij}^2}}_{\gamma}. \tag{4.28}$$

If a symmetric bias for zero motion is applied, that is $\vec{v}_{\text{ref}} = \vec{0}$, the second term ($\gamma$) on the right-hand side of (4.28) vanishes. Then, with $\sigma \to 0$, Equations (4.28) provide an infinitely close approximation of a *normal flow estimate*. Due to the bias constraint, the problem remains well-posed also in image regions where no input is present (E_x, E_y, and E_t zero). In these regions, the optical flow estimate equals the reference motion.

For $\rho > 0$ the lateral coupling between the units in the network is enabled and information and computation are spread amongst them. Now, smoothness becomes an active constraint of the optimization problem leading to a *smooth optical flow* estimate. Increasing values of ρ increase the region of support and lead to increasingly smoother flow fields (Figures 4.4b–f). Obviously, there is a clear trade-off between obtaining local motion estimates and the possibility of solving the aperture problem. A large region of support is required to solve the aperture problem but the isotropic smoothing in the resistive network cannot preserve the important optical flow discontinuities at, for instance, object boundaries.

As $\rho \to \infty$ the smoothness constraint dominates so that it does not allow the slightest spatial deviation in the flow field. The flow field becomes uniform within the whole image range and can be represented by a single *global flow* vector $\vec{v}_g = (u_g, v_g)$. The linear system (4.20) is over-complete. Analogous to the strictly local case ($\rho = 0$) the steady-state global flow solution of the network can be described in closed form as

$$u_g = \frac{\sum_{ij}(E_{yij}^2 + \sigma)\sum_{ij}(\sigma u_{\text{ref}} - E_{xij}E_{tij}) - \sum_{ij}E_{xij}E_{yij}\sum_{ij}(\sigma v_{\text{ref}} - E_{yij}E_{tij})}{\sum_{ij}(E_{yij}^2 + \sigma)\sum_{ij}(E_{xij}^2 + \sigma) - (\sum_{ij}E_{xij}E_{yij})^2}$$

$$v_g = \frac{\sum_{ij}(E_{xij}^2 + \sigma)\sum_{ij}(\sigma v_{\text{ref}} - E_{yij}E_{tij}) - \sum_{ij}E_{xij}E_{yij}\sum_{ij}(\sigma u_{\text{ref}} - E_{xij}E_{tij})}{\sum_{ij}(E_{yij}^2 + \sigma)\sum_{ij}(E_{xij}^2 + \sigma) - (\sum_{ij}E_{xij}E_{yij})^2}$$

$$\tag{4.29}$$

respectively.

In an attempt to quantify the effect of the smoothness conductances on the performance of the optical flow estimate, the network has been simulated for different values of ρ using the artificial image sequence depicted in Figure 4.4a. The estimated flow fields are shown in Figures 4.4b–f. Two different error measures were applied to quantify the error between the estimated and the exact motion field. The results are summarized in Table 4.1. In addition

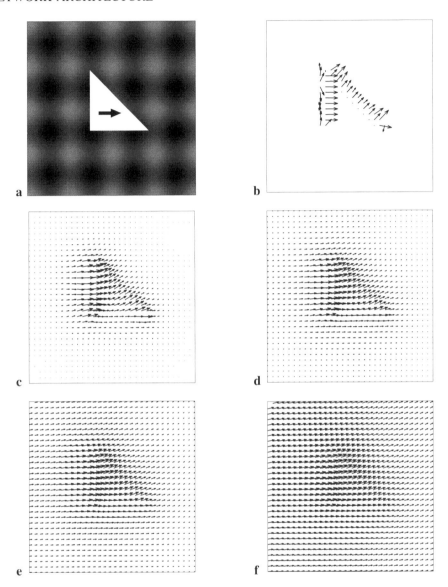

Figure 4.4 *Varying the smoothness conductance ρ.*
The network shows a wide range of computational behaviors as a function of the strength
of the smoothness conductance. (b)–(f) Keeping $\sigma = 0.001$ constant and increasing $\rho =$
$[0, 0.05, 0.15, 0.5, 1.5]$ exhibits a continuous transition from normal flow to global flow
estimation.

Table 4.1 *Estimation errors.*

The errors between the estimated and the correct image motion of the "triangle" stimulus are shown (std = standard deviation), depending on different values of the smoothness conductance ρ. Independent angular and speed errors are indicated as well as a combined error measure. Stimulus speed was normalized ($|\vec{v}_{stim}| = 1$).

Regime		Angle Φ		Speed $|\vec{v}|$		Combined Φ_c	
ρ	σ	mean	std	mean	std	mean	std
0	0.001	47.4°	29.1°	0.07	0.25	3.3°	11.3°
0.05	0.001	15.4°	9.6°	0.11	0.19	5.4°	8.9°
0.15	0.001	12.8°	7.3°	0.16	0.18	8.2°	8.5°
0.5	0.001	12.1°	5.1°	0.25	0.15	13.2°	7.2°
1.5	0.001	12.5°	3.3°	0.33	0.11	17.9°	5.3°
100	0.001	20.1°	0.3°	0.22	0.13	13.5°	4.8°
0.05	0	9.6°	5.7°	0.14	0.19	7.4°	9.1°
100	0	20.1°	0.3°	0.26	0.13	14.0°	4.6°

to the parameter values corresponding to Figures 4.4b–f, the errors were also computed for additional two parameter settings for estimating global flow ($\rho = 100$). The first error measure considers angle and speed errors independently. On one hand, the angular error decreases for increasing values of ρ because a larger region of support gradually solves the aperture problem. However, the mean angular error for the global estimates is largest because the region of support extends to the entire image space which includes two motion sources: the stationary background and the moving object. The resulting estimate represents a compromise between the background motion and the object motion. Note that a global estimate only provides the correct object motion if the background does not contain any spatio-temporal structure (see e.g. Figure 4.6). On the other hand, the average speed error increases for increasing values of ρ because smoothing assigns more and more motion to the stationary background.

A second error measure which accounts for the combined error in speed and direction is defined as

$$\Phi_c = \arccos\left(\frac{\vec{v}_{estimated} \cdot \vec{v}_{correct}}{|\vec{v}_{estimated}| \, |\vec{v}_{correct}|}\right), \tag{4.30}$$

where the optical flow vectors are considered to be the 2-D projection of the 3-D space–time direction vector $\vec{v} = (u, v, 1)$. The same measure has been used previously in a quantitative comparison of different optical flow methods [Barron et al. 1994]. Taking into account both error measures, a modest value of $\rho \approx 0.1$ provides the best results, which seems to be in good agreement with the perceptual judgment of the flow field quality (see Figures 4.4c and d). Nevertheless, such quantitative error measures have to be carefully interpreted. The performance of the network for a particular parameter setting depends significantly on the properties of the visual input.

Diffusion length

The bias constraint is crucial to keep the computational problem well-posed. Now, its influence on the optical flow estimate is considered. Figure 4.5 shows the simulated response of the optical flow network to two different image sequences. The only difference between the two sequences is the background, which is of uniform brightness in the second sequence. Figures 4.5c and d show the estimated flow fields for each sequence with parameter values $\rho = 0.1$ and $\sigma = 0$, so disabling the bias constraint, which is equivalent to the approach of Horn and Schunck [1981]. For comparison, Figures 4.5e and f show the estimates with the bias constraint enabled ($\sigma = 0.005$, with $\vec{v}_{\mathrm{ref}} = \vec{0}$). Obviously, if the bias constraint is not active, the effective extent of smoothing depends substantially on the visual input (background) and not just on the lateral conductance ρ.

To understand this effect, consider the diffusion length in a resistive network. The diffusion length L is the distance from a spatial position x_0 at which the voltage applied at x_0 has decayed to $1/e$ of its initial value. In continuous space (resistive sheet), the diffusion length is given as $L = \sqrt{\rho/\sigma}$, which is a valid approximation for the optical flow network because typically $\sigma/\rho \ll 1$ [Mead 1989]. For $\sigma \to 0$, the diffusion length becomes infinite. A single voltage applied at some location spreads across the entire network. Superposition holds because the resistive network is linear. With $\sigma = 0$, the network performs a first-order moment computation for any given voltage distribution. Returning to the example in Figure 4.5d, since the background does not provide any visual input, the units are not active and the infinite diffusion length causes spreading of the optical flow estimate over the entire image. One can argue that spreading available information into regions of no information is a sensible process [Horn and Schunck 1981]. Under some circumstances this may be the case, for example in a scene containing a single moving source. In general, however, there is little reason to assume that for regions receiving no or ambiguous visual input, the motion information from some distant image location is a better and more likely estimate than an arbitrary value. Instead, using the bias constraint provides the means to assign a priori information. It permits the network to perform better than chance by forcing the motion estimate to take on a most probable reference value. The reference motion $\vec{v}_{\mathrm{ref}}$ can be seen as the result of some long-term adaptation process that reflects the statistics of the visual environment within which the visual motion system behaves.

Reference motion

The bias constraint affects the spatial integration of visual information, necessary to estimate object motion. For a given σ, the strength of the effect depends on the contrast, the reference motion $\vec{v}_{\mathrm{ref}}$, but also on the shape of the object. In general, the resulting motion estimate deviates not only in speed but also in direction from the intersection-of-constraints (IOC) solution.

Figure 4.6 illustrates changes in the estimated global motion (4.29) as a function of decreasing figure–ground contrasts, using the same image sequence as shown in Figure 4.5b. Each of the three trajectories represents the estimates for a different reference motion. For a small σ and sufficiently high contrast, the global estimate is close to the true object motion. With decreasing contrast, however, the motion estimate becomes increasingly biased toward the reference motion. For a typical slow-motion preference (e.g. trajectory a), the direction of the estimate shifts from the IOC solution toward a normal flow vector average estimate.

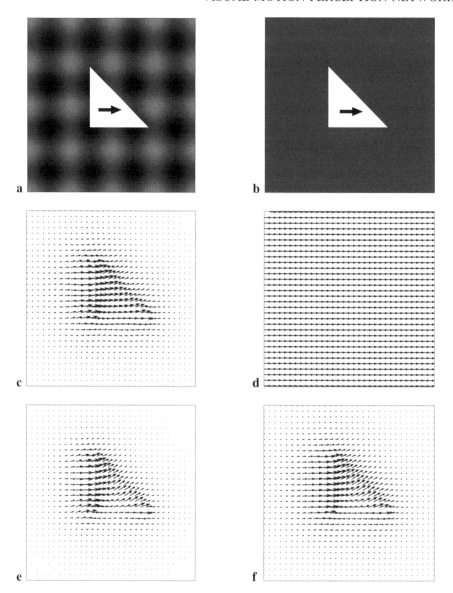

Figure 4.5 *Diffusion length.*
The same object with the same motion, but with a different background. With the bias
constraint disabled ($\rho = 0.05$, $\sigma = 0$) the behavior is extremely sensitive to the background
(c and d). Enabling the bias constraint (e and f) eliminates the sensitivity by ensuring that
the intrinsic diffusion length is finite ($\rho = 0.05$, $\sigma = 0.001$).

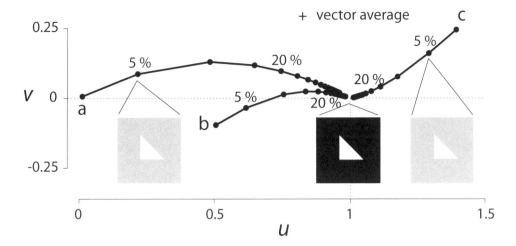

Figure 4.6 *Contrast dependence of the perceived motion.*
Global motion is estimated for various figure–ground contrasts. The correct (normalized) IOC motion of the triangle is $\vec{v} = (1, 0)$. The three trajectories represent the estimates for decreasing contrast $[95\%, \ldots, 1\%]$, each for a different reference motion $\vec{v}_{\mathrm{ref}} = [(0, 0); (0.5, -0.1); (1.4, 0.25)]$. At high contrasts, the estimates are always close to the IOC motion. At low contrasts only, the bias constraint ($\sigma = 0.001$) has a significant influence and biases the estimate toward the different reference motions. Clearly, speed as well as direction of the global motion estimate are affected. In the particular case of $\vec{v}_{\mathrm{ref}} = (0, 0)$ (trajectory a), the perceived direction of motion continuously shifts from the IOC toward a vector average estimate depending on the contrast.

Thus, for object shapes and motion for which the IOC and the vector average estimate do not have the same direction, the estimated direction of motion varies with contrast.

4.3 Simulation Results for Natural Image Sequences

Some qualitative behavior of the optical flow network is investigated using more natural image sequences. The network is simulated according to the dynamics (4.22) where the values of the conductances ρ and σ are set as indicated. More technical details about the simulation methods and the applied image sequences are listed in Appendix B.

The "tape-rolls" image sequence

The first example consists of an indoor image sequence in which two tape-rolls are rolling on a table in opposite directions. Figure 4.7 shows every second frame of the sequence, and the estimated and twice sub-sampled optical flow field provided by the optical flow network. A moderate smoothness strength ($\rho = 0.075$) was chosen, while the weight of the bias constraint ($\sigma = 0.0025$) was slightly increased compared to the previous artificial image sequences to account for the higher SNR of real image sequences. The "tape-rolls"

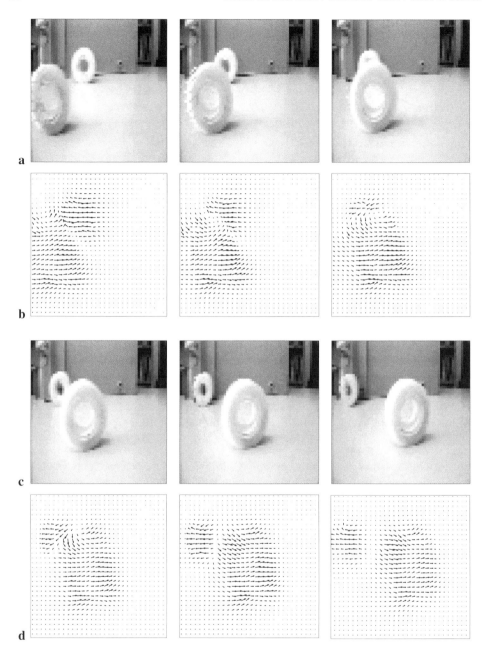

Figure 4.7 *The "tape-rolls" sequence and its optical flow estimate.*
The sequence (a,c) shows two tape-rolls rolling in opposite directions on a table. Only every second frame is shown. The estimated flow field (b,d) is down-sampled by a factor of 2. The conductances were $\rho = 0.075$ and $\sigma = 0.0025$.

sequence has many properties that makes it a challenging test for visual motion processing systems/algorithms. First of all, the spatial and temporal sampling rate is low (60×60 pixels, 15 frames/s), which increases the number of sampling artifacts. The low temporal sampling also leads to large interframe displacements (>2 pixels) which are hard to resolve. Furthermore, the sequence contains two identical objects at different spatial scales. Although the tape-rolls roll on the table, the image motion is largely translational. Still, especially for the larger roll, rotational components might influence the homogeneity of the flow sources. During several frames, strong occlusion between the two tape-rolls occurs which is difficult to resolve. For example, the brightness constancy constraint is strongly violated at occlusions. Finally, contrast conditions are not ideal, and, in conjunction with the shadows, lead to shallow brightness gradients at object boundaries.

In general, the network provides an optical flow estimate that seems appropriate. The estimate consists primarily of the translational component of the tape-rolls' motions, because there are no distinguishable visual features visible on the surface of the tape-rolls. A slight rotational component was extracted, especially at the left side of the roll in front (Figures 4.7e and f). As a consequence of the chosen weight of the smoothness constraint, the parts of the table that are visible through the holes of the tape-rolls are incorrectly estimated to move too. Furthermore, shadows associated with the rolls are perceived to move well, for example the optical flow area at the bottom of the frontal row is elongated in the direction of the shadow. The limitations of the smoothness constraint become obvious in the case where occlusion occurs and the two motion sources are present within a close neighborhood. As can be seen in Figures 4.7b–d, the partial occlusion of the distant tape-roll results in a highly disturbed flow field because the assumption that there is a single motion source is violated.

Standard sequences

On-line databases provide image sequences that can be used to test optical flow algorithms.[7] Some of these sequences have become effective standards in the literature, and permit at least some qualitative comparison among different approaches. Three of the most frequently used image sequences were applied to the optical flow network. For each sequence, a particular frame and the associated optical flow estimate are shown below. While two of the sequences are natural image recordings, the third one is artificially rendered, thus the projected motion fields are known exactly. This sequence is often used as a quantitative benchmark for different optical flow algorithms (see e.g. [Little and Verri 1989, Barron et al. 1994]). Although a quantitative error measure will be provided for reasons of completeness, I must emphasize that the significance of such numbers is fairly limited.

The first test sequence shows a well-known "Rubik's cube" placed on a platform that rotates counter-clockwise (Figure 4.8). The optical flow field correctly reflects the large rotational motion. The moderate diffusion ($\rho = 0.15$, $\sigma = 0.001$) still allows the network to resolve the local motion ambiguities at the grid pattern of the cube. This example shows that the brightness constancy constraint model, although accounting only for translational motion, may be suited to estimate large-field rotational motions too.

---

[7]See Appendix B for more details.

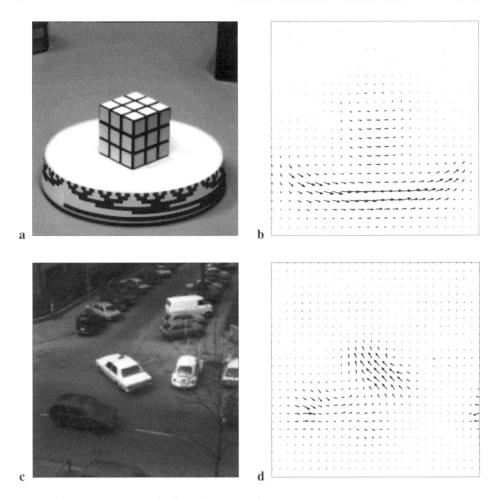

Figure 4.8 *Simulation results for other natural test sequences.*
(a,b) The fifth frame of the "Rubik's cube" sequence and the associated optical flow estimate
are shown. The result shows that large-scale rotational motion is captured well with the
translational model. (c,d) Frame #19 of the "Hamburg Taxi" sequence. Note that the low-
contrast conditions within the dark car promote a more normal flow estimate. Both flow
fields are sub-sampled by a factor of 10.

The second sequence is called the "Hamburg Taxi" sequence (Figure 4.8). It represents
a typical traffic scene and includes four different motion sources: a dark car crossing from
left to right, the taxi turning round the corner, a van passing from right to left, and finally a
pedestrian in the upper left corner walking to the left on the sidewalk. The network clearly
resolves the different motion sources, although the flow field induced by the pedestrian is
hardly visible due to its low walking speed and the sub-sampling of the flow field. The
conductance values were the same as in the previous sequence. Interestingly, the flow field
of the taxi seems more homogeneous and correct than the one of the black car in the front.

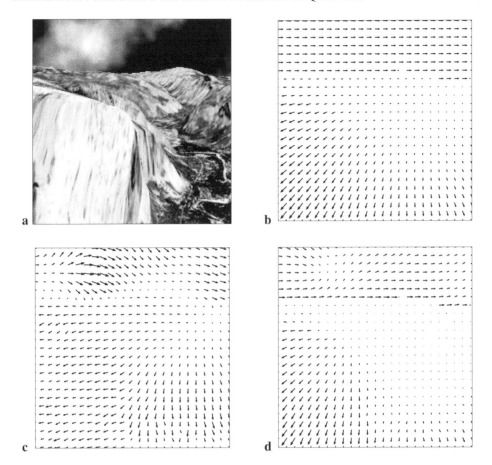

Figure 4.9 *Simulation results for the "Yosemite" sequence.*
(a) Frame #10 of the artificially rendered flight sequence over the Yosemite Valley (CA, USA). (b) The correct motion field $\vec{v}_c$. (c) The estimated optical flow field $\vec{v}_e$. (d) The difference $\vec{v}_c - \vec{v}_e$. Note especially that the optical flow estimate around the mountain peak in the foreground is affected by the aperture problem because most spatial brightness gradients in that region have essentially the same orientation. Flow fields are sub-sampled by a factor of 10.

This is mainly due to the low contrast conditions within the image region of the black car. Low contrast increases the influence of the bias constraint and results in a tendency of the flow vectors to point toward their normal direction.

The "Yosemite" sequence (Figure 4.9) represents, in addition to its artificial construction, a different motion pattern. The previous sequences were all showing moving objects relative to a static observer while this sequence shows the visual perception of a moving observer in a mostly static environment. The clouds in the sky undergo translational motion to the right. To best account for the large field motion, smoothness was assumed to be important ($\rho = 0.4$). The difference between the correct motion field and the estimated optical flow field reveals that large deviations occur around the mountain peak in the foreground (Figure 4.9d). The flow estimates around the mountain peak suffer from the dominantly vertically oriented texture in this area, which imposes the aperture problem. The flow field is biased toward the normal flow orientation which is horizontal. By contrast, the landscape in the foreground on the right-hand side shows a rich and randomly oriented texture and allows a better estimation of the flow field. A second region of large estimation errors is the cloudy sky. On the horizon, the error is large because the smoothness constraint does not preserve the motion discontinuity between the moving clouds and the mountains. Furthermore, the whole sky area has regions of large error. This is not surprising, because the successive formation and extinction of the fractal cloud pattern strongly violates the assumption of constant brightness.

A quantitative error comparison with a few previously reported methods is listed in Table 4.2. It shows that the proposed optical flow network is adequate in its performance. To provide a fair comparison, only methods that estimate a 100% dense flow field were considered. Methods that apply confidence thresholds to the flow field usually perform significantly better because they neglect regions of ambiguous or noisy visual motion where the estimation problem is particularly difficult and thus the error is likely to be high. As expected, the error is close to the one for the Horn and Schunck [1981] algorithm because the bias constraint is set rather weak and the smoothness parameter $\rho = 0.5$ is equivalent for both methods in the comparison [Barron et al. 1994]. In addition, large regions of uniform brightness where the bias constraint would be most effective are not present. Nevertheless, the finite σ seems to improve the estimation accuracy slightly by reducing the noise sensitivity in regions of low contrast. The other two methods are based on explicit matching models and provide slightly more accurate estimates for this particular sequence. All values are from [Barron et al. 1994]. Yet again, such quantitative comparison has to be interpreted cautiously because the result strongly depends on the chosen visual input.

Overall, the optical flow network provides reasonable results for realistic image sequences. For optimal results, however, the smoothness constraint must be adjusted according to the expected visual motion. For rather small moving objects in front of a stationary

Table 4.2 *Error comparison for the "Yosemite" sequence.*

The mean and the standard deviation of the combined error of the flow estimates are given for different approaches reported in the literature. To be fair, only methods that provide a 100% dense flow field are considered. The error measure is defined according to (4.30). Values are from [Barron et al. 1994].

Method	Φ_c mean	Φ_c std
Horn and Schunck [1981]	$31.7°$	$31.2°$
Anandan [1989]	$13.4°$	$15.7°$
Singh [1991]	$10.4°$	$13.9°$
The optical flow network	$23.2°$	$17.3°$

observer, ρ is preferably chosen to be small in order to obtain a reasonable spatial resolution. For large visual motions, in particular induced by ego-motion, a larger ρ results in a more robust and correct estimate.

4.4 Passive Non-linear Network Conductances

As shown, there is a strong dependence of the optical flow estimate on the values of the network conductances ρ and σ. So far, these conductances have been assumed to be constant over the entire network. The resulting isotropic diffusion process leads to uniform neighborhood interactions that are not directed by any local information of possible object boundaries. However, if such information is available, the optical flow estimate could improve significantly by making local changes to the lateral conductances such that motion integration takes place preferably over regions containing single motion sources (see Figure 2.6d). Hence, the trade-off between losing information about the locality of motion and gaining more accurate and robust estimation results through spatial integration could be circumvented. Especially, smoothing over motion discontinuities as exhibited in the previous examples may be suppressed.

In a first instance, the network conductances are considered typical two-terminal *passive* devices. The amount of current flowing through these conductances is only a (non-linear) function of the local potential differences V across them. Although called passive, the aVLSI implementation of these conductances typically consists of active circuitry. Applying passive conductances, the smoothness constraint (4.19) in the discrete case (node ij) transforms to

$$S(\Delta u_{ij}, \Delta v_{ij}) = \int_0^{\Delta^x u_{ij}} I(V)\, dV + \int_0^{\Delta^y u_{ij}} I(V)\, dV + \int_0^{\Delta^x v_{ij}} I(V)\, dV + \int_0^{\Delta^y v_{ij}} I(V)\, dV,$$

(4.31)

where $I(V)$ is the current characteristics of the passive conductance. The integrals over the currents are equivalent to the electrical co-contents [Wyatt 1995]. Global asymptotic stability is guaranteed if each integral term in the smoothness constraint (4.31) is convex with respect to V. Let $I(V) = V g(V)$ be the current through the passive conductance $g(V)$. Then, convexity is given if the second derivative of the co-content, called the *incremental conductance* g^*, is non-negative, thus

$$g^* = \frac{dI(V)}{dV} = g(V) + V\frac{dg(V)}{dV} \geq 0.$$

(4.32)

Accordingly, the optical flow network is global asymptotically stable if the incremental conductances ρ^* and σ^* at each node ij are non-negative.

What would be good functions $\rho(V)$ and $\sigma(V)$ in order to increase the performance of the optical flow estimation? The smoothness conductances $\rho(V)$ preferably should be high at places where the voltage differences V in the optical flow network and thus the variations in the flow field are small, and low if they are large. That way, small variations are smoothed out whereas large velocity gradients that typically occur at motion discontinuities are preserved. Or in other words, the smoothness constraint is modified such that it penalizes large velocity gradients less than small ones.

I distinguish two classes of implementations of such conductances. The first class describes so-called *saturating resistances* for which the *horizontal resistor (HRes)* [Mead 1989, Mahowald and Mead 1991] and the *tiny-tanh resistor* [Harris 1991] represent successful examples of appropriate electronic implementations. The current through a saturating resistance is typically a sigmoidal function of the potential V across it, thus for example

$$I(V) = I_0 \tanh(\alpha V). \tag{4.33}$$

I_0 is the maximal saturation current of the conductance and $\alpha = \rho_0/I_0$ a gain factor, where $1/\rho_0$ represents the maximal possible conductance, thus the maximal slope of the I–V curve (4.33).

Figure 4.10 shows the I–V curve, the co-content, the incremental, and the effective conductances of two sigmoidal saturating resistances characterized by (4.33). The co-content, and thus the smoothness constraint, transforms from a quadratic to a linear function of the voltage gradients. The transition point depends on the parameter ρ_0 for a given I_0: for high values (*e.g.* as in the tiny-tanh circuit), saturation occurs for very small voltages, thus the current approximates a step function, going from $-I_0$ for $V < 0$ to I_0 for $V > 0$, and the co-content becomes an absolute value function $|I_0 V|$. Absolute value functions have been suggested for solving exact constraint problems in image processing [Platt 1989]. Exact because the linear dependence of the co-content on V penalizes small deviations more strongly than in the quadratic case and thus forces the solution to be piece-wise smooth and very close to the data. Using saturating resistances, global asymptotic stability is given because the incremental conductance ρ is always non-negative.

The second class of conductances consists of so-called *resistive fuse* circuits [Harris and Koch 1989, Yu et al. 1992, Sawaji et al. 1998]. As in the case of the saturating resistances, resistive fuses allow smoothing in regions of small voltage variations across their terminals. Above some voltage difference, however, the effective conductance decreases and finally goes to zero, ending in a complete electrical separation of the two terminals. Such behavior serves well to separate distinct regions of common motion completely. The incremental conductance of a resistive fuse is partially negative (see Figure 4.10), meaning that networks of such conductances are multi-stable. Figure 4.11 compares between the behavior of the optical flow network using ohmic or passive non-linear conductances. Using saturating resistances or resistive fuses reduces significantly the smoothness of the optical flow estimate at motion discontinuities. It prevents an extensive blurring of the motion contours but still supports smoothness within areas of common motion.

Similar considerations apply for the bias conductances. The quadratic measure of the bias constraint (4.4) results in a constant conductance σ and therefore an ohmic bias current in the resistive network. The optical flow estimate is penalized in proportion to the size of its vector components. However, the original purpose of the bias constraint was to ensure that the optimization problem is well-defined in cases where the visual input is ambiguous. It seems a reasonable strategy to strengthen the bias constraint locally according to the degree of ambiguity. However, the degree of ambiguity is not at all reflected by the amplitude of the motion estimate which is the only information available for any potential passive conductance $\sigma(V)$. Therefore, it is sensible to penalize the motion estimate independently of its size. This is equivalent to a constant bias current. As shown before, such current characteristics can be approximately implemented using saturating resistances with $\rho_0 \gg 1$. Formulating the bias constraint (4.4) in terms of the appropriate electrical

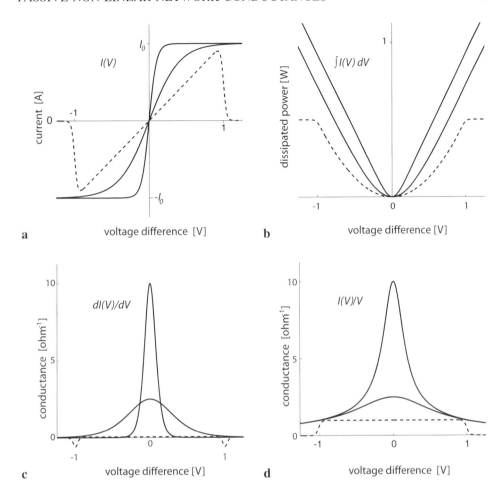

Figure 4.10 *Characteristics of passive conductances.*
The characteristic curves of two saturating resistances ($\rho_0 = [2.5, 10]$ Ω) and a resistive fuse circuit (dashed lines) are shown. (a) The current through the saturating resistances is limited for high voltage differences. The resistive fuse completely separates the two terminals at high voltage differences. This particular resistive fuse model is described by $I(V) = \lambda V\{1 + \exp[\beta(\lambda V^2 - \alpha)]\}^{-1}$ with $\lambda = 1$ Ω^{-1}, $\alpha = 1$ VA, and the free temperature-dependent parameter $\beta = 25$ [Harris et al. 1990a]. (b) The associated co-contents for the saturating resistances show the transition between the quadratic and linear characteristics as the voltage difference increases. For large ρ_0 the co-contents approximate the absolute-value function. The co-content for the resistive fuse is non-convex, which is reflected also by the partially negative incremental conductances (c) in the regions of ± 1 V. (d) Nevertheless, the effective conductances are strictly positive for both classes.

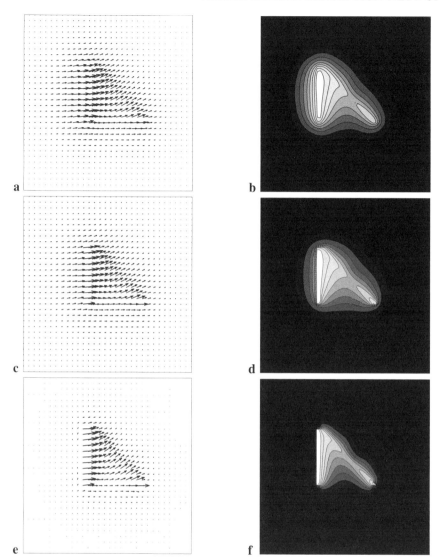

Figure 4.11 *Optical flow estimation using passive smoothness conductances.*
(a) A constant smoothness conductance (equivalent to Figure 4.4d, with $\rho = 0.15$, $\sigma = 0.001$) generates a smooth flow field, which is particularly visible in the contour plot (b) of its absolute velocities. (c,d) Saturating resistances ($\rho_0 = 0.15$, $I_0 = 0.01$) partially prevent smoothing into the background. (e,f) Resistive fuses ($\lambda = 0.15\ \Omega^{-1}$, $\beta = 200$, and $\alpha = 0.025$ VA) clearly separate the units in the network at high optical flow gradients.

co-contents leads to

$$B(u_{ij}, v_{ij}; u_{\mathrm{ref}}, v_{\mathrm{ref}}) = \int\limits_{0}^{|u_{ij}-u_{\mathrm{ref}}|} I(V)\, \mathrm{d}V + \int\limits_{0}^{|v_{ij}-v_{\mathrm{ref}}|} I(V)\, \mathrm{d}V, \qquad (4.34)$$

with $I(V)$ as in (4.33). For $\rho_0 \to \infty$, the bias constraint (4.34) reduces to the absolute-value function. Global asymptotic stability is guaranteed because the incremental conductance σ^* is non-zero for all finite ρ_0. As will be shown more quantitatively in Chapter 7, the non-linear bias conductance improves the optical flow estimate compared to a quadratic formulation of the bias constraint. The constant bias current has a *subtractive* rather than a *divisive* effect on the optical flow estimate, which is particularly advantageous in case of low-contrast and low spatial frequency visual input.

There is a general problem of applying non-linear two-terminal conductances in the optical flow network. Optical flow is a 2-D feature encoded in two resistive layers, one for each component of the flow vectors. Obviously, such conductances would regulate the connectivity in each network independently, according to local differences in the components of the flow vectors. However, it is reasonable to assume that conductance patterns should reflect the physical properties of the moving scene such as boundaries of moving objects. Since the physical properties are only dependent on the image location they should be spatially coherent within both network layers, which is not guaranteed with passive non-linear conductances. Therefore, the combined difference measure in each of the components is only an approximation of the appropriate measure for the entire feature, such as, for example, the difference in absolute vector length.

4.5 Extended Recurrent Network Architectures

Passive conductances provide limited means to control the conductance pattern in the optical flow network. In particular, passive conductances depend only on a single local scalar value, namely the voltage across their terminals. The goal is, however, to establish a more elaborate control of the network conductances such that the optical flow network is able to separate individual motion sources (objects) and estimate visual motion of these sources independently.

Controlling the lateral conductances of the optical flow network provides a very natural way to partition the image scene into disjoint regions of common motion. Disabling the conductances at the outlines of individual motion sources would electrically isolate different parts of the optical flow network, such that the collective estimate of optical flow can be restricted to take place only over individual motion sources. The control of the conductances must be dynamic in order to be able to track the moving motion sources. This concept of dynamically enabling and disabling the connections in a diffusion network is also known as *line processes* and has been proposed for different computer vision applications [Koch et al. 1986, Murray and Buxton 1987]. Each disabled smoothness conductance represents an active segment in the line process. The control of the conductances, however, does not necessarily have to be binary. While it is sensible to completely disable the network conductance at the border between two different motion sources, it can be useful to apply different degrees of smoothing in different image areas, accounting for small discrepancies between model and observation.

Line processes are an efficient concept for the optical flow estimation of multiple objects within a single network layer. The connectivity pattern serves thereby two purposes: it improves the optical flow estimate but at the same time represents the location of motion discontinuities that could be used as an input signal for later processing stages. Line processes will fail for visual scenes containing occlusions. That is, if an occluding object produces two spatially separated areas in image space that belong to the occluded object, these areas cannot be connected and treated as a single motion source if they do not connect at least in one location. Handling occlusions requires multiple network layers where each layer estimates visual motion only for one of the objects. In addition, a sophisticated mechanism is needed that decides which image locations need to be assigned to which particular layer. Such attempts have been considered using a probabilistic framework [Weiss 1997].

Also the active control of the bias conductances can be beneficial. A high local bias conductance clamps the node in the resistive layers at that location to the reference voltage $\vec{v}_{\text{ref}}$ by shunting the total node current. It prohibits the unit at that node from responding and contributing to the collective computation in the network. The unit is functionally disabled. Again, shunting does not have to be complete, depending on the strength of the bias conductance, which permits various degrees of functional suppression. Such a suppressive mechanism provides the means to instantiate selective attention. Attention is an important strategy to reduce the huge amount of visual information and focus the processing capacity only on selected and potentially interesting visual areas.

Biological systems seem to use such a strategy, which has been demonstrated in the human visual cortex where attention could account for the limited processing capacity in these areas in relation to the visual awareness of presented objects [Kastner and Ungerleider 2000]. In addition, there is physiological evidence that attentional mechanisms directly influence response properties of neurons in cortical areas related to visual motion processing. For example, it has been shown that in motion areas MT and MST of the macaque monkey, spatial attention results in an increased neural response to stimuli at the attended location and a reduced response to stimuli elsewhere [Treue and Maunsell 1996, Treue and Trujillo 1999].

To account for arbitrary conductance patterns in the optical flow network, the cost function of the optimization problem (4.5) is expressed in generic form as

$$H(\vec{v}; \rho(x, y, t), \sigma(x, y, t)) = \int_{\Omega} F + \rho(x, y, t)S + \sigma(x, y, t)B \ \ d\Omega. \tag{4.35}$$

For every given $\rho(x, y, t) \geq 0$ and $\sigma(x, y, t) > 0$ the cost function is convex and the optimization problem has a unique solution. However, global asymptotic stability is generally not guaranteed if the conductances are dependent on the optical flow estimate $\vec{v}$.

The question now is how to design appropriate mechanisms to control the conductances in the optical flow network. In the following I present a framework and discuss two particular examples to control either the smoothness conductances $\rho(x, y, t)$ or the bias conductances $\sigma(x, y, t)$. The goal is to improve optical flow estimation and to perform higher level processes such as motion segmentation and object recognition by motion.

The applied general framework is illustrated in Figure 4.12. The idea is to add a *second network* (conductance network) where each unit in that network controls one particular local conductance in the optical flow network. The network architecture is derived, again applying a constraint satisfaction approach. The network dynamics follow from gradient descent on the appropriate cost function. The conductance network and the optical flow network are recurrently connected. The output of one network is at the same time the input

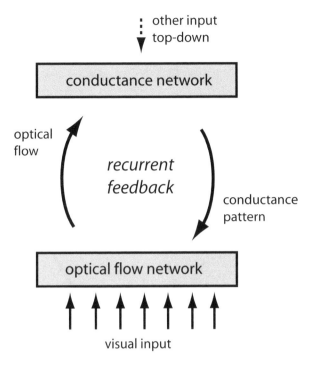

Figure 4.12 *Recurrent network architecture for the active control of the local conductances in the optical flow network.*

to the other. This recurrent processing is very powerful and yet another form of distributed processing, although crossing different levels of abstraction. Each network tries to optimize its own constraints not knowing what task the other network is performing. The powerful interplay only emerges because of the recurrent connections between the networks.

4.5.1 Motion segmentation

The first example considers the control of the smoothness conductances σ. Ideally, the conductances are such that individual motion sources are separated and the image space is segmented accordingly. To achieve this, the optical flow network is paired with a *motion discontinuity network*, consisting of two layers. Each layer is responsible only for the control of the smoothness conductances in either the x or y direction. The complete system is illustrated in Figure 4.13. Active output states P_{ij} and Q_{ij} of the discontinuity units reflect the detection of local motion discontinuities and trigger active line processes that disable the smoothness conductances of the optical flow network accordingly. Compared to passive conductances, the active control in the feedback structure permits the smoothness conductances to be set coherently in both resistive layers of the optical flow network. In order to approximate the binary characteristics of a line process, the motion discontinuity

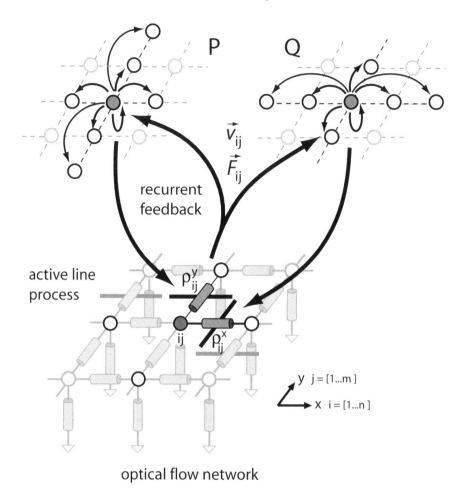

Figure 4.13 *Architecture of the motion segmentation system.*
The two layers of the motion discontinuity network control the smoothness conductances
in the optical flow network in the x and y direction respectively. A soft-WTA competition
with excitatory connections along and inhibitory connections orthogonal to the orientation
of the corresponding conductance supports the continuation of the line process by suppressing activity in the units orthogonal to the line orientation. It also adapts the threshold
of discontinuity detection by supporting a winner also in areas of low input contrast. For
reasons of clarity, the optical flow network is shown only as a single-layer network.

units[8] are assumed to have a sigmoidal activation function

$$g : p_{ij} \rightarrow P_{ij} \in [0, 1] , \qquad (4.36)$$

with a high activation gain. The smoothness conductances in the optical flow network in the y direction are directly controlled by the output P_{ij} of the discontinuity unit, thus

$$\rho_{ij}^x = \rho_0(1 - P_{ij}) , \qquad (4.37)$$

with ρ_0 being a global constant smoothness conductance.

The overall architecture and the way the output of the motion discontinuity units sets the lateral conductance pattern are defined now. Next, a description of the motion discontinuity networks is provided with the following set of constraints:

- Each active motion discontinuity unit costs.

- Motion discontinuity units are preferentially active at large motion gradients and/or locations where the brightness constancy constraint is strongly violated.

- Motion discontinuities preferentially form elongated, smooth lines.

These heuristically derived constraints can be mathematically expressed and combined as a constraint satisfaction problem of minimizing the following cost function:

$$H_P = \sum_{ij} \Big(\alpha P_{ij} + \beta(1 - P_{ij})(\Delta^y \vec{v}_{ij})^2 + \gamma(1 - P_{ij})(\Delta^y \vec{F}_{ij})^2 - \frac{\delta}{2} P_{ij}(P_{i+1,j} + P_{i-1,j})$$

$$+ \frac{\epsilon}{2} P_{ij} \sum_{w_0 \notin W} P_W + \frac{\phi}{2} \Big(\sum_W P_W - 1 \Big)^2 + 1/R \int_{1/2}^{P_{ij}} g^{-1}(\xi) \, d\xi \, \Big). \qquad (4.38)$$

The first term assigns a cost to each motion discontinuity unit that is active and thus implements the first constraint. The next two terms prefer units to be active where the coherence of motion, measured as the square of the optical flow gradient $(\Delta^y \vec{v}_{ij})^2 \equiv (\Delta^y u_{ij})^2 + (\Delta^y v_{ij})^2$ and the gradient in the y direction of the derivative of the brightness constancy constraint $(\Delta^y \vec{F}_{ij})^2 \equiv (\Delta^y F_{u_{ij}})^2 + (\Delta^y F_{v_{ij}})^2$, is disturbed. This implements the second constraint. Most known approaches for motion segmentation use only the flow field gradient as the basic measure of motion coherence [Hutchinson et al. 1988, Harris et al. 1990a, for example]. The combination of the flow field gradient with the gradient of the brightness constancy constraint has not been proposed before. The major advantage of this combined coherence measure is its robustness to the absolute level of smoothness. Figure 4.14a illustrates this for an idealized motion discontinuity. On one hand, because the brightness constancy constraint represents the deviation of the velocity estimate from the constraint line, its local gradient is largest across motion discontinuities because, there, two different motion sources collide and compete for a common flow estimate. The stronger the smoothing, the stronger the competition, and thus the larger the gradient. On the other hand, the gradient in the flow field just shows the opposite behavior. The combination

---

[8]For the sake of simplicity the discussion is limited to only one type of units (P); the analysis for the Q-units is equivalent in the x direction.

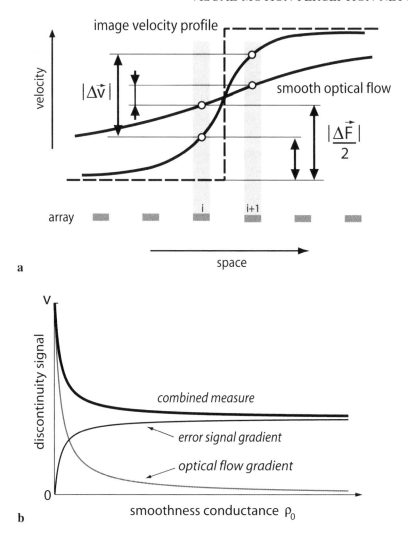

Figure 4.14 *Detection of motion discontinuities using a combined motion coherence measure.*

(a) An idealized motion discontinuity is shown with its true 1-D velocity profile (dashed line) and two examples of smooth flow estimates (bold curves). While the error signal gradient ΔF across the discontinuity is large if the flow field is smooth and small if it is not, the optical flow gradient $\Delta \vec{v}$ just shows the opposite behavior (see arrows). A combined measure thus reflects motion discontinuities rather independently of the degree of smoothness. (b) Both gradient measures are plotted as a function of smoothness. The square amplitude of the optical flow gradient $(\Delta v)^2$ and the error signal gradient $(\Delta F)^2$ are computed according to (4.21) and (4.23). Their sum is shown as the bold line. Whereas the amplitudes of the individual gradients can be zero depending on the smoothness strength, the combined measure always reflects the presence of a motion discontinuity.

of both gradient measures permits the detection of motion discontinuities rather independently of the smoothness strength. The motion discontinuities cannot "get lost" by applying large smoothness conductances. Figure 4.14b shows a more quantitative analysis for the two shaded units lying on each side of the motion discontinuity. The spatial brightness gradients at the two units were assumed to be equally large but the temporal gradient for the left unit was considered to be zero (*e.g.* a stationary background). The square norms of the optical flow gradient $(\Delta \vec{v})^2$ and the error signal gradient $(\Delta \vec{F})^2$, and their sum, were computed and plotted as a function of the lateral smoothness conductance ρ_0. There is a lower bound for the combined measure which is identical to the error signal gradient for $\rho_0 \to \infty$. Thus, ideally, by applying an appropriate threshold the motion discontinuity can be detected no matter how smooth the flow field is, which is not true if using either one of the gradient measures alone. The combined measure is maximal when the smoothness conductance between the units is zero. This is exactly the case when a line segment is active and disconnects the two units.

The first three terms in (4.38) do not yet define a network structure because they create no dependency between individual units of the motion discontinuity network. This situation changes with the next term (δ): it favors those units to be active that have active nearest neighbors in the direction of the orientation of their line segments. Activity is spread along the orientation of the line segments to promote continuous smooth motion discontinuities. The next two terms support this process by inducing a weighted WTA competition (weighted by the kernel W) along the units perpendicular to the orientation of the line segment, which suppresses the formation of many, very close, and parallel-oriented line segments. The constraints for the weighted WTA competition are formulated identically to the classical WTA example (3.8), except that competition is now spatially weighted by the (1-D) neighborhood kernel W. The weights are chosen to decay exponentially with distance and are normalized such that their sum is equal to one. The kernel is assumed to be symmetric where w_0 is its central coefficient. The weighted WTA competition also provides the means for an adaptive activation threshold because it promotes a winner also in regions of low input contrast. The danger of forcing discontinuities even when no input is present is counter-balanced by the first two constraints. The last term in (4.39) is necessary due to the finite gain of the activation function g of the discontinuity units. Although chosen here to be a typical sigmoidal function (3.7), g can be any function that is non-linear, monotonically increasing, and limited to [0, 1].

The cost function (4.38) is bounded from below for any given and finite inputs $(\Delta \vec{v})^2$ and $(\Delta \vec{F})^2$. Thus, a network that performs gradient descent on the cost function (4.38) is taken to be asymptotically stable. This leads to the following dynamics for the motion discontinuity networks:

$$\dot{p}_{ij} = -\frac{1}{C} \left[\underbrace{\alpha - \beta(\Delta^y \vec{v})^2 - \gamma(\Delta^y \vec{F})^2}_{\text{static threshold}} \right.$$

$$\left. \underbrace{- \delta(P_{i+1,j} + P_{i-1,j}) + \epsilon \, \overline{P_{ij}} + \phi \, \overline{\overline{P_{ij}}} - \epsilon \, w_0 P_{ij} - \phi}_{\text{soft-WTA with cross-excitation}} + \frac{p_{ij}}{R} \right] \qquad (4.39)$$

where $\overline{P_{ij}} = \sum_W w_W P_W$ stands for the weighted average of the output activity within the neighborhood W and $\overline{\overline{P_{ij}}} = \sum_W w_W \overline{P_W}$ for its average. The first three terms in (4.39) represent a static threshold function where the unit P_{ij} is turned on if the sum of the weighted input is larger than a constant α. The network architecture follows directly from the dynamics (4.39). Similar to the WTA network shown in Figure 3.3, connectivity can be reduced by introducing two layers of additional inhibitory units. The first layer receives weighted input from all units within a single neighborhood and thus provides $\overline{P_{ij}}$. The second layer does the same but now receives input from the first layer and therefore provides $\overline{\overline{P_{ij}}}$. Besides inhibitory input, each unit also gets excitatory input from different sources: a general and constant input ϕ, input from its neighbors in the orientation of its line segment, and self-excitation according to the parameters ϵ and w_0, respectively.

Global asymptotic stability for the discontinuity networks is difficult to test. It is the connection weight matrix of the network and thus the parameters δ, ϵ, ϕ, and W that decide its stability properties. Strict convexity of the cost function can be easily verified for the (simple) classical WTA network by requiring the Hessian of the cost function to be positive definite (see Chapter 3). However, such a procedure becomes complex for arbitrary neighborhood kernels W and may only be solved numerically. In order to get some estimate of what the conditions are for strict convexity and thus for global asymptotic stability an eigenvector analysis was performed for two extreme cases of W that allow analytical solutions:

- $W = [1]$
  The kernel includes only the local unit itself. The excitatory and inhibitory ϵ terms in (4.39) cancel and the smallest eigenvalue leads to the sufficient condition

$$\frac{p_0}{R} + \phi - \sqrt{2}\delta \geq 0 \qquad (4.40)$$

  for global asymptotic stability, where $1/p_0$ is the maximal gain of the activation function g. Clearly, since the excitatory and inhibitory ϵ terms in (4.39) cancel, only the nearest-neighbor excitation remains(δ term). It has to be smaller than the inhibition imposed by the leak (driving force) p_0/R and the (self-)inhibition ϕ.

- $W = [w_0, \ldots, w_N]$, where $w_i = 1/N \quad \forall \; i \in [1, N]$
  The kernel now includes a complete row (or column) of N units with uniform weights. This changes the characteristics to a classical WTA competition along the particular row or column respectively. Therefore, the eigenvalue condition

$$\frac{p_0}{R} - \frac{\epsilon}{N} - \sqrt{2}\delta \geq 0 \qquad (4.41)$$

  is equivalent to the self-excitation gain limit (Equation (3.14) on page 39) including also excitation from the nearest neighbors. The row length N appears in (4.41) since the weights are normalized.

General kernels will lead to more complex conditions. Numerical experiments, however, suggest that condition (4.41) under the rigorous assumption $N = 1$ represents a conservative lower limit for which global asymptotic stability holds. Increasing the inhibitory weight parameter ϕ always exhibits a positive effect on this limit, so permitting stronger excitatory connections ϵ and δ.

In order to close the recurrent loop, the smoothness conductances in the optical flow network have to be replaced by the appropriate expression (4.37). Therefore, the original smoothness constraint of the optical flow problem is replaced with

$$S = \sum_{ij} \rho_0 \left[(1 - P_{ij}) \left((\Delta^y u_{ij})^2 + (\Delta^y v_{ij})^2 \right) + (1 - Q_{ij}) \left((\Delta^x u_{ij})^2 + (\Delta^x v_{ij})^2 \right) \right],$$

(4.42)

where $\Delta^x u_{ij}$, $\Delta^x v_{ij}$ and $\Delta^y u_{ij}$, $\Delta^y v_{ij}$ are the linearized gradients of the optical flow field in the x and y directions respectively. The effective conductance is within the range $[0 \ldots \rho_0]$.

The complete system (discontinuity networks plus optical flow network) is asymptotically stable because each sub-network is. However, global asymptotic stability is difficult to investigate. It is clear that the interaction between the optical flow and the discontinuity networks is non-linear due to the multiplicative interaction between the different state variables as seen in Equation (4.42). Any statements on additive networks are therefore not applicable. Qualitatively, the combined system is expected to be multi-stable because the feedback loop certainly has a positive component.

Simulations

The complete recurrent system was simulated for the "tape-rolls" sequence. The parameters for the two motion discontinuity networks were set as follows: the static threshold $\alpha = 0.12$, the input weights $\beta = \gamma = 1$, neighbor excitation $\delta = 0.002$, the soft-WTA parameter $\epsilon = 0.003$, and $\phi = 0.13$. Self-inhibition was set rather small with p_0/R, resp. $q_0/R = 0.01$. Clearly, the parameter values obey the excitation gain limit for the classical WTA case (4.41). The bias conductance and the reference motion of the optical flow network were kept as in previous simulations of the optical flow network alone ($\sigma = 0.0025$, $\vec{v}_{\text{ref}} = \vec{0}$). Figures 4.15a and c show every second frame of the sequence, overlaid with the combined activity in the two layers of the motion discontinuity network. The associated optical flow estimate is shown below each frame superimposed on a gray-scale image of its absolute value (Figures 4.15b and d).

The assessment of the simulation results must remain on a qualitative and descriptive level because absolute benchmarks are not available. The overall performance is reasonable, especially given the fact that there was no applied reset of the network states in between different frame presentations. A reset clearly would help to improve the segmentation quality because it eludes the problem of hysteresis. Unfortunately, most of the reported approaches in the literature do not show any simulation results of whole image sequences with continuous frame presentation [Hutchinson et al. 1988, Harris et al. 1990b, Memin and Perez 1998, Aubert et al. 1999, Cesmeli and Wang 2000, for example]. Thus, it seems possible that these results are based on single-frame simulations with default initial conditions. It is therefore difficult to judge how strongly these approaches are susceptible to hysteresis. It is clear that if an interframe reset is applied in the presented discontinuity networks, hysteresis is of no concern. Most likely, a reset would allow more optimal settings, which would lead to better single-frame results than shown here.

Hysteresis is noticeable, for example, when comparing the first frame in Figure 4.15 with all subsequent frames: since the first frame is not burdened by any history, it provides a much cleaner estimation of the motion discontinuities. In later frames, the estimated discontinuities appear to correlate less with the physical boundaries of the objects, which is especially obvious within the large tape-roll. Spurious discontinuities cause an inhomogeneous flow estimate within the large tape-roll. Nevertheless, the coarse outline detection of the two

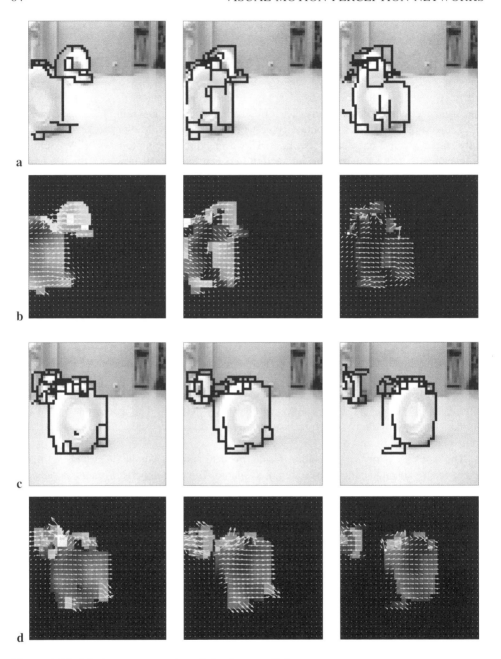

Figure 4.15 *Motion segmentation of the "tape-rolls" sequence.*
(a,c) Every second frame of the sequence is shown overlaid with the output activity of the
motion discontinuity network. (b,d) The associated optical flow is displayed as a vector
field and gray-scale image of its absolute value.

objects is good and does not show much hysteretic artifacts. This is surprising regarding, for example, the low-contrast conditions at the trailing edge of the tape-roll in front. Only in the last frame does part of the trailing boundary seem to get stuck, which causes the disrupted contour.

As expected from the limitations of the brightness constancy constraint, occlusions cause some trouble and lead to small partitions of unreliable motion estimates. It is interesting also to see that the system cannot distinguish between moving objects and dynamic shadows which can be observed in part near the bottom of the tape-roll in front.

4.5.2   Attention and motion selection

The second example considers the control of the bias conductances in the optical flow network. The *motion selective network* is a single-layer network extension to the optical flow network. As sketched in Figure 4.16, units of this network are retinotopically organized, receive input from the associated units of the optical flow network in the form of the optical flow estimate, and feed back their output to control the bias conductances. The task of the motion selection network is to control the bias conductances in the optical flow network such that the optical flow estimation is limited only to spatial areas that correspond to moving objects (motion sources) matching a characteristic expected motion $\vec{v}_{model}$. This imposes the following constraints on the units in the motion selective network:

- Activity in the motion selective network is restricted to only sub-parts of the image (attentional focus).

- Active units cluster and form a contiguous area.

- Only those units are active which correspond to image locations where the optical flow estimate matches some expected motion $\vec{v}_{model}$.

Again, these constraints can be formalized as a constraint satisfaction problem, minimizing the following cost function:

$$H_A = \frac{\alpha}{2} \sum_{ij} \sum_{kl \neq ij} A_{ij} A_{kl} + \frac{\beta}{2} \left(\sum_{ij} A_{ij} - N_{max} \right)^2 - \frac{\gamma}{2} \sum_{ij} \sum_{W} A_{ij} A_W$$

$$- \delta \sum_{ij} A_{ij} (\vec{v}_{ij} \cdot \vec{v}_{model}) + 1/R \sum_{ij} \int_{1/2}^{A_{ij}} g^{-1}(\xi) \, d\xi, \tag{4.43}$$

where A_{ij} stands for the output of an individual motion selective unit at node (i, j) and $g : a_{ij} \rightarrow A_{ij} \in [0, 1]$ is its saturating activation function.

The first two terms are the equivalent constraints of the simple WTA network (3.8) with the small difference that the total activity is forced to be N_{max}. By choosing a saturating activation function that is monotonic and limited on the interval $[0, 1]$ such a constraint will allow ideally N_{max} winners. These first two terms define a *multi-winners-take-all* (mWTA) network. In contrast to the weighted WTA circuit of the motion discontinuity networks, every unit in the array participates in the competition and has an equal chance of winning. The fixed number of winners can be seen as the maximal amount of attentional resources

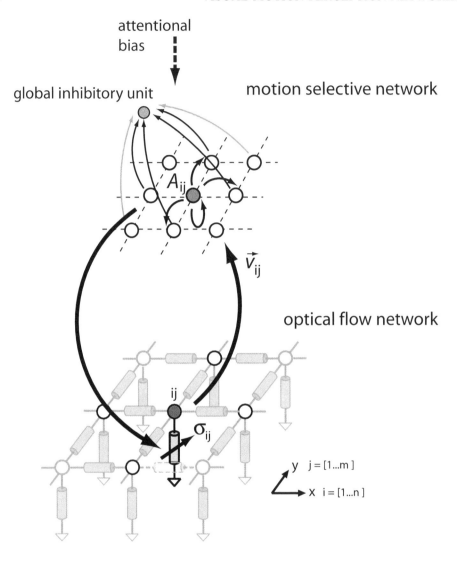

Figure 4.16 *Architecture of the motion selective system.*
The optical flow network is extended with the motion selective network which is selective for a particular type of motion. The network performs a multi-WTA competition that
promotes a cluster of winning units rather than only a single winner. The output of the
motion selective network is fed back to control the strength of the local bias conductances
in such a way that optical flow estimation is selectively restricted to image areas matching
a particular kind of motion, specified by $\vec{v}_{model}$.

for which the input must compete. Such a paradigm of limited resources has been proposed
as the underlying mechanism accounting for attentional processes in primates [Kastner
and Ungerleider 2000]. Saturation of the activation function is crucial because, otherwise,
the desired multi-winner behavior would not occur, and a single active unit with output

$A_{winner} = N_{max}$ would correspond to the optimal solution of (4.43). The activation function was chosen to be sigmoidal (3.7).

The third term favors units to be active within an active neighborhood W, hence it promotes the network to form clusters of activity. This term is important in order to form spatial coherent areas of activation that represent the extents of the selected motion sources. The fourth term represents the input to the network and prevents units from being active at locations where the optical flow estimate does not match the expected motion. The chosen dot-product is just one possible difference measure. Finally, the analog units with activation functions of finite slope require some activation cost, which is expressed in the last term.

Applying gradient descent on the cost function (4.43) leads to the following dynamics of the network:

$$\dot{a}_{ij} = -\frac{1}{C}\left[\frac{a_{ij}}{R} + \alpha \sum_{kl \neq ij} A_{kl} + \beta\left(\sum_{ij} A_{ij} - N_{max}\right) - \gamma \sum_W A_W - \delta(\vec{v}_{ij} \cdot \vec{v}_{model})\right]$$

$$= -\frac{1}{C}\left[\frac{a_{ij}}{R} + (\alpha + \beta)\sum_{ij} A_{ij} - \alpha A_{ij} - \beta N_{max} - \gamma \sum_W A_W - \delta(\vec{v}_{ij} \cdot \vec{v}_{model})\right]. \quad (4.44)$$

The network architecture is defined by these dynamics and is very similar to the one of the classical WTA network (compare Figure 3.3 on page 39): each unit receives global inhibitory input from all units in the network $(\alpha + \beta)$ but also excites itself (α). In addition, each unit has excitatory connections (γ) from and to units in a defined neighborhood W. Finally, a constant excitatory input which is proportional to N_{max} is applied to all units. Again, the connectivity is massively reduced using a global inhibitory unit that sums up the output of all units and inhibits each one accordingly.

The cost function (4.43) is bounded from below and above for any given and finite flow field $\vec{v}$ and is a Lyapunov function of the system (4.44). To assure global asymptotic stability within the mWTA network, the self-excitation gain limit has to be obeyed. Again, this limit depends on the exact form of the excitation kernel W. In the simple case of a four nearest-neighbor kernel with equal weights, the condition

$$\alpha + \gamma \leq \frac{a_0}{R} \quad (4.45)$$

is sufficient to guarantee global asymptotic stability of the motion selective network. $1/a_0$ is the maximal gain of the activation function, and R is the leak resistance. For more general kernels the appropriate conditions have to be found numerically.

Note that the cost function (4.43) is a soft representation of the imposed constraints. Thus the total number of winners in the network is not necessarily N_{max}. In fact, as seen in the classic WTA network, the maximal loop gain can be too small to determine clear winners for weak and/or ambiguous input. The total loop gain, if necessary however, can be increased almost arbitrarily using cascades of mWTA networks as proposed earlier (see also Figure 3.4 on page 41).

Closing the recurrent feedback loop between the motion selective network and the optical flow network allows the motion selective units to control the strength of the local bias conductances. Hence, the bias constraint in (4.19) is modified to

$$B = \sum_{ij}(\sigma_1 + (1 - A_{ij})\,\sigma_2)\left[(u_{ij} - u_{ref})^2 + (v_{ij} - v_{ref})^2\right], \quad (4.46)$$

with $0 < \sigma_1 \ll \sigma_2$. Using a typical sigmoidal activation function as in (3.7) $A_{ij} \in [0, 1]$, thus the effective bias conductances are in the interval $[\sigma_1, \ \sigma_1 + \sigma_2]$.

Simulations

The motion selective system was simulated twice for the "tape-rolls" sequence, once for preferred rightward motion ($\vec{v}_{model} = (1, 0)$ [pixel/s]) and once for leftward motion ($\vec{v}_{model} = (-1, 0)$ [pixel/s]). The results are displayed in Figure 4.17 and Figure 4.18 respectively. Every second frame of the sequence is shown, superimposed with the resulting optical flow field. Below each frame, the corresponding activity in the motion selective network is indicated. A sufficient activation of optical flow units at selected visual locations requires a high gain in the WTA competition ($1 - A_{ij} \to 0$). This is achieved using a two-stage mWTA cascade, leading to the almost binary output activity pattern of the motion selective network. The network parameters were kept constant in both cases and set as follows: the mWTA parameters $\alpha = 0.02$, $\beta = 0.5$, and $\gamma = 0.05$, the input weight $\delta = 1$, and self-inhibition $a_0/R = 0.1$. The default number of winners was slightly reduced in the preferred leftward motion case to account for the size difference between the two object (the two tape-rolls). The smoothness conductance and the lower limit of the bias conductance in the optical flow network were kept as in the smooth optical flow example (see Figure 4.7) and the maximal leak conductance was set as $\sigma_2 = 0.8$.

For preferred rightward motion, the optical flow estimate for the smaller roll in the back rolling to the left is almost completely suppressed while the flow field of the front roll is correctly estimated. The activity in the motion selective network approximately matches the size of the tape-roll and does not show any shift toward the trailing edge which would be expected as the result of hysteresis. Preferring leftward motion (Figure 4.18), the system assigns all its attentional activity to the small roll in the back. As before, the optical flow field everywhere outside the attentional focus is suppressed. All parameters besides the preferred motion vector were kept identical in both cases. Again, there was no reset of the network states in between frames.

Stability of the complete system is guaranteed because each sub-system, the optical flow network and the motion selective network, is asymptotically stable.[9] Although a detailed analysis of conditions for global asymptotic stability of the complete system is not performed, it seems very likely that the positive feedback can elicit hysteresis. In fact, hysteresis is observed in the simulation results: for example, in Figure 4.17, where the top part of the tape-roll in front is not, or is only partially, selected although its motion matches the preferred motion. This is likely to be initiated by the occlusion of the two tape-rolls occurring in the first few frames. Hysteresis prevents the network from becoming active again right after occlusion. Instead, the network needs some time to overcome the "experience" before it recovers. However, reapplying the whole sequence but deleting the network history after occlusion (resetting the network states after frame #8) leads to an immediate recovery of the activity pattern in the top part of the tape-roll (not shown). Although hysteresis occurs, it seems to be less pronounced and distorting than in the previously discussed segmentation network.

---

[9]The time constants of both networks are assumed to be much shorter than the time-scale on which visual motion occurs.

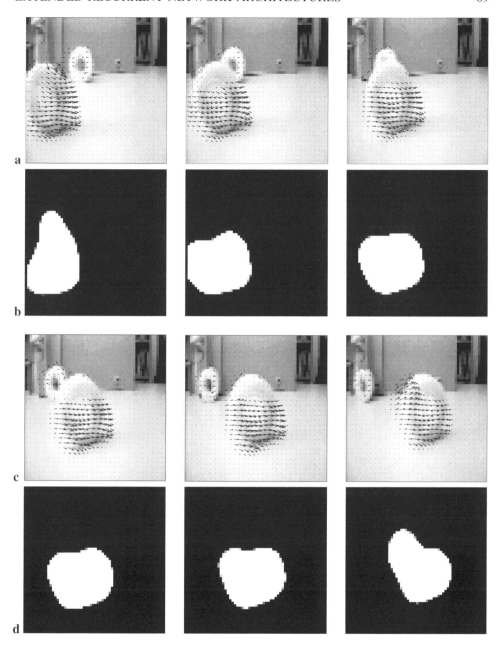

Figure 4.17 *Motion selection for preferred rightward motion.*
(a,c) The "tape-rolls" sequence (frame # 2,4,6,8,10,12) is shown overlaid with the resulting optical flow estimate. The attentional feedback clearly suppresses the flow field associated with the smaller tape-roll. (b,d) The activity in the motion selective network ($N_{max} = 500$) is shown for the associated frames. Note that no reset of the network states was performed between different frames.

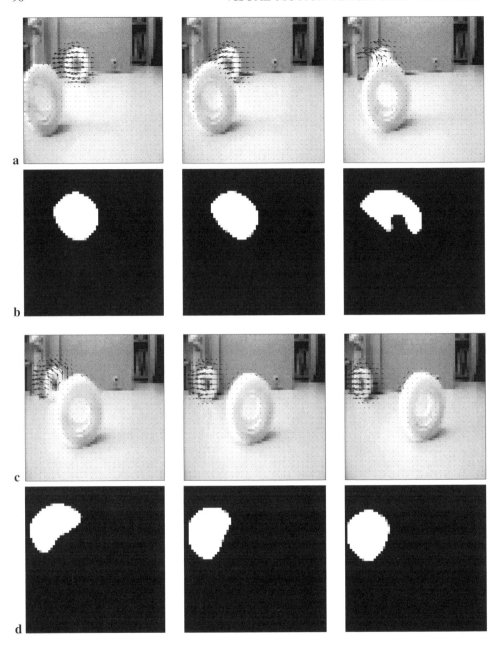

Figure 4.18 *Motion selection for preferred leftward motion.*
(a,c) The same "tape-rolls" sequence, but now leftward motion is favored. (b,d) The associated activity in the motion selective network is clearly assigned to the smaller roll. The size of attention is slightly reduced in comparison to the previous example ($N_{max} = 300$).

The motion selective network provides the means for top-down input. First, it contains two global parameters, the preferred motion $\vec{v}_{model}$ and the size N_{max} of the attentional kernel. These two parameters could be controlled by the output of another, yet even higher level network. The parameters clearly represent object-related properties such as size and motion that are *independent of a retinotopic representation*. Networks that provide such output will inevitably follow a different, object-oriented encoding scheme. It is interesting to note that these object-related inputs are imposed simply as a global leak (N_{max}) or a filter operation ($\vec{v}_{model}$), as shown in the dynamics (4.44). Second, top-down input could be directly provided to the motion selective units. In terms of the formulated constraint satisfaction problem, such attentional input I_{ij} can be easily incorporated by adding a term $\epsilon \sum_{ij}(1 - A_{ij})I_{ij}$ to the cost function (4.43). Depending on the conductance strength ϵ, the input could impose only a weak attentional bias, or it could lead to a strongly guided attentional focus (attentional spotlight). Presumably, the input I_{ij} represents the output activity of another retinotopically arranged place-encoding network or some lower dimensional, variable-encoding map [Hahnloser et al. 1999].

4.6 Remarks

In this chapter, I have shown how optical flow estimation can be formulated as a constraint satisfaction problem. The bias constraint thereby turns out to be important and reveals many interesting properties. As shown, it is the addition of this constraint that transforms the estimation of visual motion into a well-posed problem under any visual input conditions. The bias constraint permits the application of a priori knowledge about the expected visual motion which becomes important if the available visual information is ambiguous or even absent.

More important, I have demonstrated how the constraint satisfaction problem can be solved by an analog electronic network. The basic optical flow network consists of two cross-connected resistive layers of nearest-neighbor connected processing units. According to the strengths of the network conductances, the resulting output broadly varies between a normal flow and a global flow estimate. Changing the conductance strength permits – to some extent – programming of the optical flow network.

Two extensions have been discussed that each use an additional network. These extensions perform motion segmentation and motion selective attention tasks where the activity in the additional networks directly represents motion discontinuities or the attentional strength, respectively. Since segmentation and attention are decisive processes by definition, different variations of the classical WTA architecture form the core circuits of these additional network layers. In particular, a cross-excitation weighted WTA architecture is introduced that forces the formation of continuous motion boundaries while suppressing nearby parallel-oriented line segments. A multi-WTA architecture is proposed for the motion selective attention network that assigns a preferred number of winning units to the input area that corresponds best to a preferred motion. An important aspect of decisive processes within the temporal domain is their possible multi-stable behavior that can produce hysteretic effects. The networks are designed so that they exhibit as little hysteresis as possible.

This is achieved by respecting the weight conditions for mono-stability in the decisive network layers (self-excitation gain limits). Furthermore, the positive loop gain is reduced by minimizing the feedback effect on the inputs to the decisive stages (combined gradient differences and non-complete suppression). Simulation results using real-world image sequences demonstrate that these arrangements keep hysteresis on an acceptable level.

For all proposed networks, highest priority was given to simple network architectures that permit a physical implementation as integrated electronic circuits.

5

Analog VLSI Implementation

The previous chapter showed how to define optimization problems that describe the computational process of visual motion estimation and how to design appropriate analog networks that solve them. This and the next two chapters now show ways how to actually *build* such networks as analog integrated circuits. Foremost, I will describe in this chapter the circuit diagram of the optical flow unit, the basic unit of all following network implementations, and discuss its individual building blocks and their properties.

This is a rather technical chapter. The intention was to describe and discuss all building blocks and sub-circuits of the optical flow unit in detail so that the technically interested reader who has access to the necessary technology and tools can, in principle, reproduce the circuits. Or better, adapt and improve them in new designs. Those readers who are not familiar with integrated circuits will find a short primer on basic transistor models and simple circuits in Appendix C. It should be sufficient for a basic understanding of the circuits presented in this chapter. Clearly, a thorough comprehension of the circuit details requires a more profound background. The interested reader is referred to a list of excellent reference books at the end of Appendix C. Although quite technical, the chapter addresses strongly the computational consequences of the chosen way of implementation. The focus of description is on those circuits that have a strong influence on the computational outcome of the systems. In particular, the impact of non-linearities will be discussed thoroughly.

5.1 Implementation Substrate

In the 1970s when Kunihiko Fukushima and his colleagues reported a first analog electronic model of the outer-plexiform layer of a typical vertebrate retina, they presented two man-high laboratory instrumentation racks housing the 700 photodiodes and thousands of discrete operational amplifiers and capacitors that their implementation required. And to display the retina's behavior another one of these racks was needed, containing 700 electric light-bulbs(!) whose intensity reflected the activity of the emulated retinal ganglion cells [Fukushima et al. 1970]. It is very likely that at that time, Fukushima and his colleagues did not have access to integrated semiconductor technology. So, due to its obvious

Analog VLSI Circuits for the Perception of Visual Motion   A. A. Stocker
© 2006 John Wiley & Sons, Ltd

lack in efficiency – just consider the power consumption – their "artificial retina" had to remain rather a proof of concept than a useful device. Luckily, today we are in a much different position. Integrated circuits are found everywhere in products of daily life. Driven mainly by the digital circuits industry, integrated circuits technology did and still does advance tremendously and at a fast pace (*e.g.* Moore's law [Moore 1965]). And while the development of this technology is tailored to digital circuits, it is up to the designer to decide how to operate these circuits. As Carver Mead said:[1] "all circuits are analog, even if they are used as digital building blocks". Integrated circuits provide the means to build efficient visual motion systems. The high integration density allows us to create complete systems on a single chip that can be applied on behaving autonomous agents with tight efficiency constraints. Exploiting the transistor properties in the *sub-threshold* regime (weak inversion) leads to the design of analog circuits that operate in the nanoampere range, providing low-power solutions even for network architectures of substantial size.

Silicon *complementary metal oxide–semiconductor (CMOS)* technology is the technology of choice for the implementations of the optical flow network and its variations discussed here.[2] Modern CMOS technology has many advantages. Due to its widespread use it is highly evolved and therefore relatively cheap and available. It provides two complementary transistor types (hence the name) which make it possible to design compact and sophisticated circuits. Polysilicon capacitive layers provide the means to create high-density capacitors, which is a basis for temporal signal processing. And some parasitic devices such as lateral bipolar transistors provide the circuit designer with some additional freedom for innovative design. A big advantage of CMOS technology is that phototransduction circuits can be incorporated seamlessly without any extra processing steps. Complete *system-on-chip* solutions can be built where phototransduction and signal processing are combined within a single chip. Furthermore, the topology of the nearest-neighbor connected networks can be preserved in the silicon implementation, allowing the design of so-called *focal-plane* arrays.

Figure 5.1 shows a microphotograph of the smooth optical flow chip discussed in more detail in the next chapter. Clearly, it exemplarily reveals the modular array structure of identical units, typical of such a focal-plane processor. The chosen AMS 0.8 μm process[3] is conservative compared to current processes that have a minimal feature size one order of magnitude smaller. The choice of the particular process was guided mainly by convenience. It must be emphasized that all of the presented chip implementations are prototypes; they are considered proofs of concepts rather than products. Nevertheless, they have been used in some applications. The circuits are not tailored specifically to the AMS process. The same circuit architectures could be ported to more modern processes with smaller feature size. In principle, focal-plane architectures permit us to choose an arbitrary array size, only constrained by the available silicon real-estate. The processing power scales proportionally with the array size and thus the processing speed remains practically constant. The following table gives an idea of how these analog array processors compare to digital microprocessors.

---

[1] In his foreword to *Analog VLSI: Circuits and Principles* [Liu et al. 2002].

[2] As shown later, the necessity to implement a pair of ideal diodes did actually require a BiCMOS process instead.

[3] For exact specifications see Appendix D.

Figure 5.1 *Analog, distributed, and time-continuous computation in silicon.*
A magnified picture of the *smooth optical flow chip* (Chapter 6) with a network size of
30 × 30 optical flow units is shown. Each unit of the network is identical and contains all
the necessary circuitry for focal-plane processing, including phototransduction.

It lists some physical properties of a modern Pentium-4 processor and the smooth optical
flow chip shown in Figure 5.1.

What Fukushima and his colleagues clearly realized from building their analog retina
model was that wires are costly in physical implementations. A pure feedforward connection
scheme would have required a prohibitively large number of wires to implement the center-
surround filter characteristics of their retina model. Instead, they used two resistive networks
to account for the inhibitory and excitatory kernels, and by simply subtracting the output
of corresponding nodes they achieved the desired spatial filtering operation.

5.2 Phototransduction

The first step in visual motion processing is the transduction of the visual information
into appropriate electronic signals. Being the first stage in the processing stream, the

Table 5.1 Comparing the physical properties of a Pentium-4 processor and the smooth optical flow chip (www.intel.com and
www.pcstats.com).

	Smooth optical flow chip	Pentium-4 (Prescott)
fabrication	AustriaMicroSystems AMS	Intel
clock frequency	continuous time	3.4 GHz
min. feature size	0.8 μm	0.09 μm
no. of transistors	$\approx$100,000	$\approx$125,000,000
die size	21.2 mm^2	112 mm^2
power dissipation	0.05 W	103 W (max)

characteristics of the phototransduction stage crucially affect subsequent processing. The
operational limits of the phototransduction stage define the maximal possible operational
range for the whole system.

On one hand, the demands on the phototransduction stage are imposed by the characteristics of the visual environment within which the system is expected to operate. Some
systems actively control their visual environment by providing a well-defined artificial illumination of the scene, which is possible for a small and isolated environment. A particularly
good example is the optical computer mouse,[4] which uses an optical motion sensor to measure its motion. The space between the mouse and the surface it glides on is small and
well shielded from external light. So a light-emitting diode (LED) is sufficient to control
the illumination using a reasonably small amount of power. A freely behaving autonomous
system, however, cannot rely on such a constrained environment, nor can it actively control
the illumination with a reasonable amount of resources. Rather, it has to operate on its given
visual environment. For a typical real-world environment, this means that the system has to
be able to operate over a very wide range of light intensities, yet at the same time it must
remain sensitive to small contrast variations. These two properties are difficult to combine
for a phototransduction stage because they impose opposing constraints: phototransduction
gain has to be small in order to respond to the wide dynamic range whereas high gain is
needed to resolve small contrast variations. Biology solved this problem. The human retina,
for example, responds over a dynamic range of more than *ten* orders of light intensities and
still is able to resolve contrast differences of less than 1% [Foley and Legge 1981, Bird
et al. 2002]. These are remarkable specifications.

On the other hand, the particular structure of the visual motion perception networks also
constrains the phototransduction stage. It is sensible, for example, to keep the continuous-
time processing in order to exploit the advantage of not facing temporal aliasing during
motion estimation.

5.2.1 Logarithmic adaptive photoreceptor

The logarithmic adaptive photoreceptor is a circuit concept for phototransduction that meets
the above requirements on bandwidth, sensitivity, and continuous-time operation. It is an
example of how a biologically neural principle can be successfully translated into an analog

---

[4]Sensor: Agilent (www.agilent.com); product, e.g. Logitech (www.logitech.com).

electronic circuit. Many circuit variations of this concept have been realized [Delbrück and Mead 1989, Mahowald 1991, Delbrück 1993b] and its application in the "Silicon Retina" [Mahowald and Mead 1991] is still one of the most striking examples of its ability to mimic biology. The logarithmic adaptive photoreceptor applies two computational key principles to achieve a wide dynamic range and high sensitivity: *logarithmic compression* and *adaptation*. In essence, it encodes the light intensity into a logarithmic voltage signal. Yet, it has a low-gain response for slowly changing light intensities whereas it responds with high gain for transients. Adaptation by means of a feedback loop ensures that the high-gain response range, thus the operating point of the photoreceptor, is always adjusted according to the slow dynamics of the light intensity signal.

Figure 5.2a shows the circuit diagram of the logarithmic adaptive photoreceptor circuit as it is used in the implementations of the visual motion networks presented here. It follows the design of Delbrück and Mead [1994] with an adaptive element as proposed by Liu [1998]. Phototransduction is initiated in the photodiode, a reverse-biased pn junction, where incoming photons generate electron–hole pairs, and thus a photocurrent $I_{ph}(t)$ that is proportional to the light intensity. Without the feedback loop, the photodiode and the feedback transistor M_1 form a simple source-follower photoreceptor. The voltage across the photodiode logarithmically decreases with increasing $I_{ph}(t)$ for a fixed gate voltage $V_f(t)$ because the typical photocurrent is within the sub-threshold regime of M_1. The feedback loop adjusts the gate voltage $V_f(t)$ of the feedback transistor such that the voltage across the photodiode is basically clamped. Now, the voltage $V_f(t)$ is logarithmically encoding the photocurrent $I_{ph}(t)$. Feedback thereby acts on two different pathways: the first path is through the adaptive element, thus via transistor M_2. This path accounts for the steady-state response of the photoreceptor. At steady state, the photoreceptor output voltage $E(t)$ ideally equals $V_f(t)$, thus

$$E(t) \approx V_f(t) = \frac{V_T}{\kappa} \log(I_{ph}(t)/I_0) + c_0 \qquad \text{(steady state)}, \qquad (5.1)$$

where $V_T = kT/q$ is the thermal voltage (approximately 25 mV at room temperature), I_0 is the sub-threshold transistor model scaling factor, and c_0 is a constant voltage determined by the amplifier bias voltage Bias_{Ph}. The second feedback path is through capacitive coupling via C_1. This path dominates the transient response of the photoreceptor because at the onset of a transient in the photocurrent the conductance of the adaptive element is low. The feedback gain from the output $E(t)$ onto $V_f(t)$ is determined by the capacitive divider formed by C_1 and C_2, thus

$$E(t) \approx \frac{C_1 + C_2}{C_1} V_f(t) \qquad \text{(transient response)}, \qquad (5.2)$$

with $V_f(t)$ as in (5.1). With the appropriate choice of C_1 and C_2 the transient gain can easily be one order of magnitude or more. Figure 5.2b shows the response of the adaptive photoreceptor to a set of transient step-edge stimuli of different contrasts, measured under different mean illumination over several orders of magnitude. It clearly illustrates the gain difference between the slowly changing mean and the transient signal. The signal amplitude of the transient response is approximately constant due to the logarithmic encoding. The photoreceptor output is basically encoding stimulus contrast.

Equations (5.1) and (5.2) describe the response of the photoreceptor at the two ends of its dynamic spectrum, where adaptation either did not start (transient) or was already

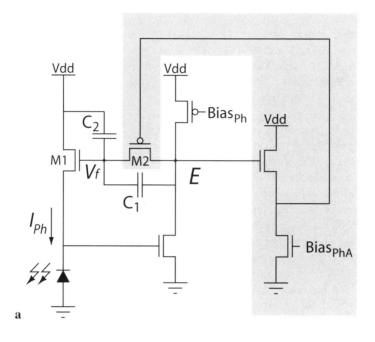

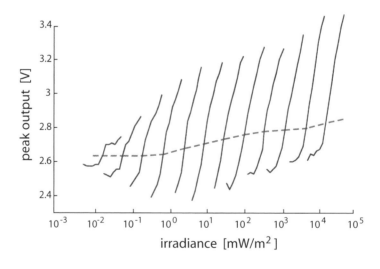

Figure 5.2 *Adaptive photoreceptor.*
(a) Circuit diagram of the adaptive photoreceptor circuit with adjustable adaptation time constant. The shaded area outlines the three transistors that form the adaptive element. (b) Transient responses of the photoreceptor to an on/off step-edge stimulus of many different contrast levels measured at different mean irradiances. While the steady-state response approximately increases logarithmically with irradiance (dashed line), the transient response is approximately independent of it. The circuit operates over more than five orders of irradiance levels. Adapted from [Delbrück and Mead 1994].

completed (steady state). Between these two points, the response is determined by the adaptation behavior, hence by the characteristics of the adaptive element. The adaptive element consists of a well transistor whose gate is controlled by a native-type source follower that follows the photoreceptor output $E(t)$ with a gain factor $\kappa < 1$ and an offset given by the voltage Bias_{PhA}. At steady state, M_2 is out of saturation and the conductance is approximately zero. Following an onset transient and now in saturation, the adaptation current through M_2 is to first order exponential in the voltage difference between $E(t)$ and the gate voltage of M_2. Because of the non-unity gain in the follower, this difference grows linearly $(1 - \kappa)$ with $E(t)$ and thus ensures that the adaptation current is exponential in the transient amplitude as long as M_2 is in saturation. This implies that the adaptation time constant is approximately uniform. For an offset transient, a similar scenario holds where the adaptation current is exponential in $E(t)$, this time with an exponential gain of κ. The bias voltage Bias_{PhA} permits us to control the dynamics of adaptation by setting the current level where the exponential dependency begins once M_2 is in saturation. The back-gate effect makes this level also dependent on the mean output voltage of the photoreceptor, hence on the bias voltage Bias_{Ph}.

The logarithmic adaptive photoreceptor acts as a temporal band-pass filter where the characteristic frequencies are adjustable to some degree. Its high-pass characteristics are due to the amplification of transient signals. The corner frequency is somewhat adjustable by increasing or decreasing the bias voltage of the adaptive element. Similarly, its low-pass characteristics depend on the output amplifier and the cut-off frequency can be adjusted to some degree with the bias voltage Bias_{Ph}. A band-pass filter stage at the very beginning of the visual motion process is a good way to discard non-relevant high- and low-frequency contents of the visual information (such as the absolute intensity level or flicker noise) from propagating into the following processing stages.

5.2.2   Robust brightness constancy constraint

Brightness $E(t)$ defines a subjective measure of intensity and is determined by the characteristics of the phototransduction stage. Unlike traditional computer vision, perceptual processing does not start after the imaging process. Phototransduction is the first processing step for visual motion estimation and it is part of the entire perceptual system. How does adaptation and logarithmic compression influence visual motion estimation of the optical flow network?

The main effect of adaptation is that it assigns a different gain to moving and static visual input. For moving stimuli, the photoreceptor predominantly operates in its high-gain transient regime because adaptation typically takes place on a much longer time-scale than the observed intensity changes. For static input the photoreceptor is adapted and the gain is low. The gain is directly reflected in the magnitude of the spatiotemporal brightness gradients. As we discussed, the strength of the feedback currents (4.23) at each node of the optical flow network is proportional to the local spatial brightness gradients. Therefore, predominantly only those network units are active where visual motion is present. Units receiving static visual input such as in the background do hardly contribute because their photoreceptors are adapted. Thus, the adaptive photoreceptor already performs some simple form of figure–ground segregation by assigning low-response gain to static image areas!

The logarithmic encoding of intensity has another positive influence on the robustness of the optical flow estimation. According to (5.1), the spatio-temporal brightness gradients can be generally written as

$$\frac{\partial}{\partial \chi} E(\chi) \propto \frac{1}{I(\chi)} \frac{\partial}{\partial \chi} I(\chi) \tag{5.3}$$

where $I(\chi)$ is the intensity and χ is any of its space–time directions. Clearly, the brightness gradients are proportional to the normalized intensity gradients, which is nothing more than *contrast*. The input to the optical flow network becomes largely invariant on absolute intensity. Contrast is an object property while intensity depends on both the object and the environment (illumination). Thus, the logarithmic adaptive photoreceptor not only acts as a phototransduction stage but also preprocesses the visual information in a way that leads to a more robust and more invariant implementation of the brightness constancy constraint. It reduces the impact of static visual information on the collective computation of visual motion and makes it independent of the illumination in the visual environment.

5.3 Extraction of the Spatio-temporal Brightness Gradients

The spatio-temporal brightness gradients have to be computed from the output of the photoreceptor circuit. This section introduces circuits that allow us to compute the spatial and temporal derivatives of the voltage signal $E_{ij}(t)$ at each node of the optical flow network. It discusses their limitations and subsequent influence on the motion estimation process, in particular the consequences of spatial sampling of the brightness distribution.

5.3.1 Temporal derivative circuits

Ideal temporal differentiation of a voltage signal can be performed by a single capacitor, where the current I_C flowing into the capacitance C is proportional to the temporal derivative of the voltage V_C across it, thus

$$I_C = \frac{1}{C} \frac{dV_C}{dt}. \tag{5.4}$$

Typically, in real circuit applications the capacitive current I_C is measured with a linear resistor R that is connected in series with the capacitance C. In addition, a high-gain feedback loop is needed to provide a high-impedance input. It guarantees that the voltage across the capacitance remains as close as possible to the input signal and thus differentiation is correct, despite the resistor R.

Figure 5.3 shows two possible configurations for such a differentiator circuit. The *grounded-capacitor* circuit (Figure 5.3a) forces the potential V_C on the capacitive node to follow the input signal V_{in}. The output voltage V_{out} represents a superposition of the input signal and its temporal derivative which becomes evident from the circuit's transfer function

$$H_a(s) = \frac{1 + \tau s}{1 + A^{-1}(1 + \tau s)} \approx 1 + \tau s \tag{5.5}$$

where A is the gain of the amplifier. The *clamped-capacitor* differentiator (Figure 5.3b), however, clamps the capacitive node V_C to the reference potential V_{ref}. The voltage V_{out} is

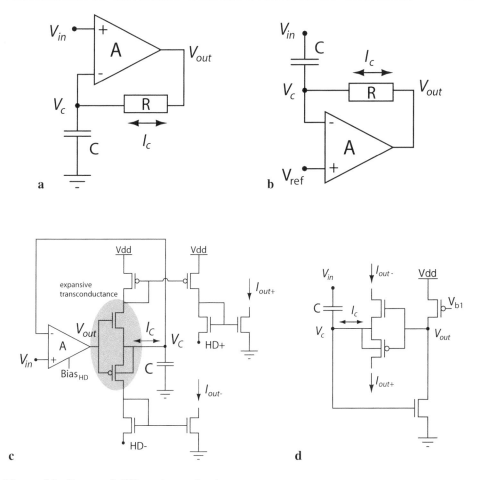

Figure 5.3 *Temporal differentiator circuits.*
Two basic configurations are possible using high-gain feedback circuits: (a) the *grounded-capacitor* differentiator, and (b) the *clamped-capacitor* differentiator. (c) The modified hysteretic differentiator circuit [Kramer, Sarpeshkar, and Koch 1997] provides a rectified output current that is ideally proportional to the temporal derivative of the input voltage V_{in}. (d) The implementation of a clamped-capacitor differentiator requires only one capacitance and four transistors [Stocker 2003]. In contrast to the hysteretic differentiator it avoids any dependence on the DC-level of the input signal.

directly (negatively) proportional to the temporal derivative of the input signal as expressed by the transfer function

$$H_b(s) = -\frac{\tau s}{1 + A^{-1}(1 + \tau s)} \approx -\tau s. \tag{5.6}$$

In both circuits, I_C represents the temporal derivative of the input signal V_{in} for ideal amplification and linear R and C. The approximations are valid if the feedback gain is high, thus $A \gg 1$.

CMOS implementations of both types of differentiator circuits have been developed and are widely applied. Figure 5.3c shows the circuit diagram of a grounded-capacitor differentiator circuit, which was originally called a "hysteretic differentiator" [Mead 1989]. Amplifier A needs to be fully differential (*e.g.* a five-transistor operational transconductance amplifier)[5]. An expansive transconductance element replaces the linear resistor R in the feedback loop [Kramer et al. 1997]. It is an active circuit consisting of a pair of transistors of opposite polarity, which are connected at their sources. This element delivers the necessary capacitive current I_C. Depending on the direction of I_C, current flows either through the upper transistor (nFET) to charge the capacitance C or through the lower transistor (pFET) to discharge C. The current through each branch is mirrored with current mirrors to provide rectified versions I_{out+} and I_{out-} of the capacitive current. Variable source potentials $HD+$ and $HD-$ permit the currents to be scaled.

Figure 5.3d shows the circuit diagram of a compact clamped-capacitor differentiator [Stocker 2003]. It uses the same expansive transconductance element as the hysteretic differentiator. The feedback amplifier, however, is not required to be fully differential. A two-transistor inverting amplifier is sufficient and makes the design compact. Again, current mirrors must be added to read out the rectified currents I_{out+} and I_{out-} (left out in the circuit diagram). Unlike in the ideal linear circuits (5.5) and (5.6), the voltage difference $(V_{out} - V_C)$ in both implementations no longer represents a valid approximation for the derivative of the input voltage. Due to the expansive non-linear characteristics of the transconductance element, the output voltage approximates an almost binary signal.

The performance of both circuits is very similar, yet in the following, only data from the clamped-capacitor circuit (Figure 5.3d) are presented. Figure 5.4 shows the measured response for two different input signal waveforms. The rectified output currents nicely represent the positive and negative derivatives of the input signal. Form and amplitude (see also Figure 5.5a) are close to ideal. Small deviations occur at sign changes of the input derivatives. Because the amplifier gain is finite, there exists a small signal range ΔV_{in} where $\Delta(V_{out} - V_C)$ is small enough to remain in the low-transconductance regime of the expansive transconductance element. The transconductance is too low to provide a sufficiently high capacitive current I_C such that the voltage V_C is not able to follow the input (grounded-capacitor differentiator), and I_C is too small to clamp V_C (clamped-capacitor differentiator). There is a computational advantage in this behavior, such that low-amplitude noise signals are suppressed.

The accuracy of differentiation is illustrated by the measured peak currents as a function of the maximal temporal gradient of a sinewave input signal (Figure 5.5a). For a slight above-threshold biasing of the amplifier, the circuits can truthfully differentiate signal gradients up to 10^4 V/s. This wide dynamic range is another advantage of the expansive transconductance element. The amplifier does not have to provide the capacitive current I_C. Its bandwidth is only limited by its output conductance and the parasitic capacitances at the output node. In sub-threshold operation, the output conductance of the amplifier is linear in the bias current. Above threshold, this turns into a square-root dependency. Figure 5.5b illustrates this by showing the cut-off frequency (3 dB attenuation) for a sinewave input signal as a function of the bias voltage of the amplifier. The semi-logarithmic representation directly illustrates the linear dependency on the bias current. Above threshold (at

---

[5]See Appendix C.3.

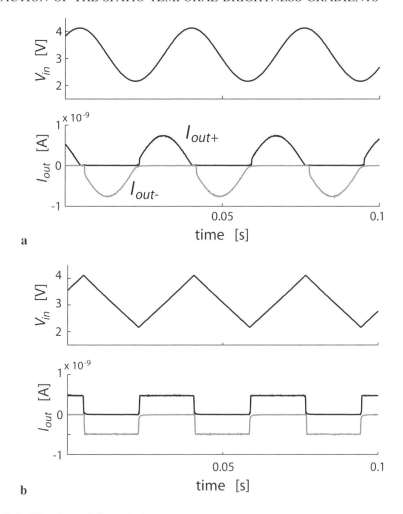

Figure 5.4 *Waveform differentiation.*
Measured response characteristics of the clamped-capacitor circuit (Figure 5.3b) for (a) a sinewave and (b) a sawtooth input signal of equal temporal frequency and amplitude. The rectified output currents approximate well the waveform of the derivatives of the input signals and also the amplitudes are correct, with the peak of the sinewave derivative being $\pi/2$ times the amplitude of the sawtooth derivative.

approximately 0.8 V) the curve starts to deviate from the linear fit. The measurements also indicate that for the typical temporal frequency spectrum of a real-world environment, it is sufficient to bias the amplifier well below threshold.

Although both circuits behave very similarly, the advantage of the clamped-capacitor differentiator is its compact design. Also, because the feedback voltage V_C is clamped and independent of the input signal level, any DC dependence is naturally eliminated, whereas this is not the case for the hysteretic differentiator. While computing the derivative of

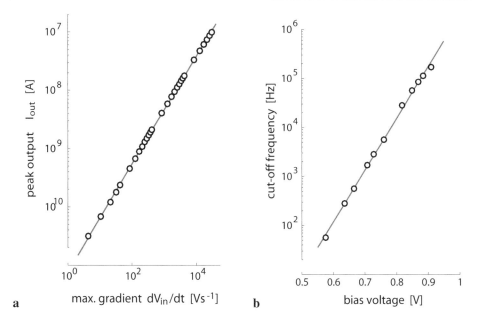

a max. gradient dV_{in}/dt [Vs^{-1}] **b** bias voltage [V]

Figure 5.5 *Linearity and bandwidth.*
(a) Measured peak output current amplitudes of the clamped-capacitor differentiator, shown as a function of the applied temporal voltage gradient for a sinewave input (peak-to-peak amplitude 2 V). (b) The cut-off frequency (3 dB amplitude attenuation) as a function of the amplifier bias voltage. Note that the linear dependency on the bias current only holds in the sub-threshold regime.

the photoreceptor signal, however, this is not of concern. In the context of a focal-plane implementation of the optical flow network the size advantage is certainly more important. Nevertheless, the reader will later notice that the implementation of the optical flow unit still contains the hysteretic differentiator. This is simply because the clamped-capacitor circuit was not invented at the time the optical flow network circuits were developed. Future implementations will profit from using this more compact circuit.

5.3.2 Spatial sampling

While the photoreceptor circuit permits phototransduction to be continuous in time, it necessarily remains discrete in space. The photoreceptor output in a focal-plane array inevitably represents a spatial sample of the image brightness distribution. Consequently, visual information is lost, leading to a classical reconstruction problem as dealt within sampling theory. While the continuous-time output of the photoreceptor prevents temporal aliasing,[6] spatial aliasing can occur. As a consequence, the estimation of the continuous spatial gradient based on the sampled brightness distribution is an ill-posed problem [Bertero et al. 1987].

---

[6]Definition: *aliasing*–folding of higher into lower frequency components in a discrete spectrum due to under-sampling of the signal.

Figure 5.6a illustrates the situation for a one-dimensional array of optical flow units with grid constant d. Each unit contains a photoreceptor circuit with a photodiode of size D. Visual information is spatially sampled and locally averaged over D. The fraction of transduced light is indicated by the so-called *fill-factor*, defined as the ratio $\delta = D/d$. Clearly, for any 2-D active pixel sensor $\delta < 1$, and it usually decreases as the computational complexity of a single pixel increases. A stimulus is presented to this simplified model of the optical flow network such that it generates a sinusoidal brightness pattern at the photoreceptor output. The stimulus moves with a fixed velocity v whereas spatial frequency can vary. Thus, the brightness distribution is given as

$$E(x,t) = \sin(kx - \omega t) \qquad \text{with } x, t \in \mathbb{R}, \tag{5.7}$$

where $\omega = kv$ is its temporal and k the spatial frequency in units of the Nyquist frequency $[\pi/d]$. Three sinewave stimuli with different spatial frequencies are shown in Figure 5.6a. Using one of the temporal differentiator circuits provides an almost perfect temporal differentiation of the photoreceptor signal. Hence, taking into account the photodiode size D, the temporal gradient of the brightness at location x_i becomes

$$
\begin{aligned}
E_t(x_i, t) &= \frac{\partial}{\partial t} \frac{1}{D} \int_{x_i - D/2}^{x_i + D/2} \sin(k(\xi - vt)) \, d\xi \\
&= -\frac{v}{D}[\sin(k(x_i - vt + D/2)) - \sin(k(x_i - vt - D/2))] \\
&= -\frac{2v}{D}[\cos(k(x_i - vt)) \cdot \sin(kD/2)]. \tag{5.8}
\end{aligned}
$$

The estimation of the spatial brightness gradient $E_x(x_i, t)$ requires that the signal be reconstructed from the sampled values in order to define its gradient. The 2-D problem is not trivial. There are elaborate methods to define appropriate filters such that important properties (*e.g.* rotational invariance) are preserved [Farid and Simoncelli 2003]. A focal-plane implementation, however, is constrained by the limited silicon real-estate available. Solutions that do not need extensive wiring are favored. For this reason, a simple model for the spatial gradient is chosen. It assumes that the brightness distribution is linear in between two sampling points. Thus, the local spatial gradient at a sampling point is defined as the central difference operator. Using this model, the continuous 1-D spatial gradient at location x_i is defined as

$$
\begin{aligned}
E_x(x_i, t) &= \frac{1}{2d}(E(x_{i+1}, t) - E(x_{i-1}, t)) \\
&= \frac{1}{2d}\left(\frac{1}{D}\int_{x_i + d - D/2}^{x_i + d + D/2} \sin(k(\xi - vt)) \, d\xi - \frac{1}{D}\int_{x_i - d - D/2}^{x_i - d + D/2} \sin(k(\xi - vt)) \, d\xi\right) \\
&= \frac{1}{kDd}\left[\sin(k((x_i + d) - vt)) \cdot \sin(kD/2) - \sin(k((x_i - d) - vt)) \cdot \sin(kD/2)\right] \\
&= \frac{2}{kDd}\sin(kd) \cdot \sin(kD/2) \cdot \cos(k(x_i - vt)). \tag{5.9}
\end{aligned}
$$

Figure 5.6b shows the computed spatio-temporal brightness gradients at zero phase $(t = 0, x_i = 0)$ according to (5.8) and (5.9) for a sinewave grating pattern moving with

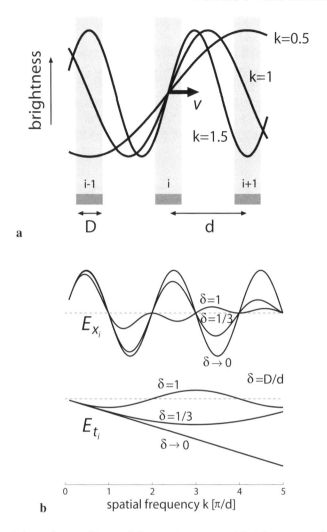

Figure 5.6 *Spatial patch sampling and the spatio-temporal brightness gradients.*
(a) Spatial sampling of sinusoidal brightness patterns of different spatial frequencies k (given in units of the Nyquist frequency $[\pi/d]$) in one visual dimension. The dark shaded areas represent the photosensitive area D of the photoreceptors, over which the brightness distribution is locally integrated. Aliasing occurs at frequencies $k \geq 1$. (b) Computed spatial and temporal gradients according to (5.8) and (5.9), at zero phase ($t = 0$, $x_i = 0$). An increasing photodiode size affects both gradients by introducing low-pass filter characteristics. Even the sign can change. However, these effects are only significant for high fill-factors at high spatial frequencies that are typically not relevant in practice.

speed v. For spatial frequencies $k < 0.5$ both gradients approximate well the true continuous derivatives. For higher frequencies, the estimate of the spatial gradient increasingly deviates from the true value and changes sign at multiples of the Nyquist frequency. The temporal derivative estimate is not affected in the same way, thus a cancellation of the

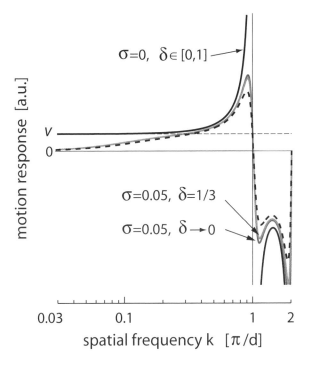

Figure 5.7 *Optical flow estimate as a function of spatial stimulus frequency.*
The expected response of a single optical flow unit according to (5.10) is shown as a
function of the spatial stimulus frequency k for a given speed v. Solid curves are the
peak responses ($x = 0$; $t = 0$) while the dashed curve is the signal averaged over an entire
stimulus cycle ($\sigma = 0.05$, $\delta = 1/3$). Only for low spatial frequencies does the motion
response resemble the true velocity. For $k > 0.5$ the linear approximation of the spatial
gradient leads to a significant increase in response amplitude. At multiples of the Nyquist
frequency spatial aliasing reverses the direction of the perceived motion. Without the bias
constraint ($\sigma = 0$) the computation is ill-posed at multiples of the Nyquist frequency and the
response goes to $\pm\infty$ in their vicinity. Increasing the fill-factor does *not* affect the motion
response unless $\sigma > 0$, yet the influence remains small. A finite σ biases the response at
very low frequencies to the reference motion $V_{\text{ref}} = 0$.

effect must not be expected (see Figure 5.7). It is important to mention that the fill-
factor δ does not have a significant effect for frequencies that are actually relevant in
practice.

Having defined the spatio-temporal gradients on the spatially discrete but temporally
continuous brightness distribution, we can investigate the effects of stimulus and array
parameters on the estimate of visual motion. For simplicity, consider a single optical flow
unit with no lateral connections. In equilibrium, it is expected to report normal flow (4.28)
that in the 1-D case reduces to

$$V_{out}(x_i, t) = -\frac{E_t(x_i, t)E_x(x_i, t) + V_{\text{ref}}\sigma}{\sigma + E_x(x_i, t)^2}. \tag{5.10}$$

Substituting (5.8) and (5.9) into (5.10), the output of a single optical flow unit can be quantified as a function of spatial frequency of the stimulus, the fill-factor δ, and the weight of the bias constraint σ. The reference motion is set to zero ($V_{\text{ref}} = 0$) to simplify the illustration. Figure 5.7 shows the predicted peak output for different parameter values as a function of the spatial stimulus frequency k. Note that the response depends strongly on k. Only for low spatial frequencies does the motion output approximate well the correct velocity. For very low frequencies, the locally perceived stimulus contrast diminishes and the non-zero σ biases the output response toward the reference motion. As the spatial frequency approaches the Nyquist frequency, the response increases to unrealistically high values and changes sign at $k \in \mathbb{Z}^+[\pi/d]$. A non-zero σ and an increasing fill-factor reduce this effect. However, the influence of the fill-factor is only marginal for reasonably small values of σ.

In the extreme and hypothetical case where $\sigma = 0$, (5.10) reduces to

$$V_{out} = v \; \frac{kd}{\sin(kd)} \qquad \text{with } u_{\text{ref}} = 0. \tag{5.11}$$

The computation is ill-posed at spatial frequencies $k \in \mathbb{Z}^+$ and the motion response approaches infinity close to their vicinity. For small k, however, the response approximates well the true velocity. Two things are remarkable: first, D drops out of Equation (5.11), which means that low-pass filtering by averaging over D has no influence for $\sigma = 0$. Second, the response depends neither on time nor space, meaning that the periodic sinewave stimulus leads to a constant output signal.

However, this changes for $\sigma > 0$. According to (5.10), the response is now given as

$$V_{out}(x_i, t) = v \; \frac{kd}{\sin(kd)} \cdot \frac{\cos(kx_i - \omega t)^2}{\alpha + \cos(kx_i - \omega t)^2} \tag{5.12}$$

with

$$\alpha = \sigma \; \frac{k^2 d^2 D^2}{4 \sin^2(kd) \sin^2(kD/2)}.$$

Since $\alpha > 0$, the motion response $V_{out}(x_i, t)$ remains in the range between zero and the hypothetical constant response given by (5.11) at any time t. The exact value depends on the spatial frequency k of the stimulus and the parameters d and D. Furthermore, the motion response is now *phase-dependent* due to the remaining frequency terms. The response follows the stimulus frequency ω and thus oscillates sinusoidally. As a consequence, the time-averaged motion response over the duration of an entire stimulus cycle (dashed curve in Figure 5.7)

$$\overline{V}_{out}(x_i) = \int_t^{t+T} V_{out}(x_i, t) \, dt \quad \text{with } T = \frac{1}{\omega} \tag{5.13}$$

is always less than the peak response.

The phase dependence of the motion response is an inherent property of the non-zero σ and the single spatial sampling scale in the spatial brightness gradient estimation. A similar problem is encountered in motion energy approaches, where quadrature pairs of local spatio-temporal filters with a spatial phase shift of $\pi/2$ have to be combined in order to achieve phase-independent motion energy responses [Adelson and Bergen 1985, Grzywacz and Yuille 1990, Heeger 1987a]. However, coupling between units in the optical flow network reduces local phase dependence.

5.4 Single Optical Flow Unit

Figure 5.8 shows the complete circuit diagram of a single unit of the optical flow network, including the adaptive photoreceptor and the differentiator circuit. The diagram consists of two identical parts for each of the two optical flow components, including the negative feedback loops necessary to embed the brightness constancy constraint. The general encoding scheme is differential: that is, each variable is encoded as the difference of two electrical signals, whether it is a current or a voltage. For example, the two estimated components of the optical flow vector are given as the potential differences $(U_+ - V_0)$ and $(V_+ - V_0)$, where V_0 is a set virtual ground.

Although all labels are coherent, the analogy to the idealized architecture (Figure 4.2) is recognized easiest when considering the two capacitive nodes C_u and C_v. The current sum on these nodes has to be zero according to Kirchhoff's law. The sum consists of four individual currents, namely

- the feedback current I_F of the recurrent loop $(Fu_- - Fu_+)$ and $(Fv_- - Fv_+)$ respectively,

- the output current I_B of the transconductance amplifier,

- the current I_S flowing to neighboring units in the resistive network, and finally

- the capacitive current I_C.

The only purpose of the circuit is to make sure that these currents match the functional description given by the dynamics (4.22). Specifically the following should hold, illustrated only for the u component of the optical flow:

$$I_F \equiv (Fu_- - Fu_+) \propto -E_{x_{ij}} (E_{x_{ij}} u_{ij} + E_{y_{ij}} v_{ij} + E_{t_{ij}})$$

$$I_S \propto \rho (u_{i+1,j} + u_{i-1,j} + u_{i,j+1} + u_{i,j-1} - 4u_{ij})$$

$$I_B \propto \sigma (u_{ij} - u_{\text{ref}}). \tag{5.14}$$

The next few paragraphs will discuss how good the match is and what consequences deviations have on the quality of the estimated optical flow.

Most expensive, in terms of the number of transistors, is the implementation of the feedback loops (indicated by circular arrows in Figure 5.8). It requires the emulation of several mathematical operations (5.14). Two wide-linear-range Gilbert multipliers provide the four-quadrant multiplication $E_x u$ and $E_y v$ between the spatial gradients and the instantaneous estimate of the optical flow. Summation of the multipliers' output and the output currents of the temporal derivative circuit is performed at the pair of current mirrors. The summed currents are then mirrored to a second pair of Gilbert multipliers that perform the outer multiplication with the spatial brightness gradients. Finally, the loop is closed by cascode current mirrors that reconnect the multipliers' outputs to the capacitive nodes C_u and C_v respectively.

5.4.1 Wide-linear-range multiplier

The design of the multiplier circuit needed to compute $E_x u$ and $E_y v$ is critical for a successful a VLSI implementation of the optical flow network. The specifications for this

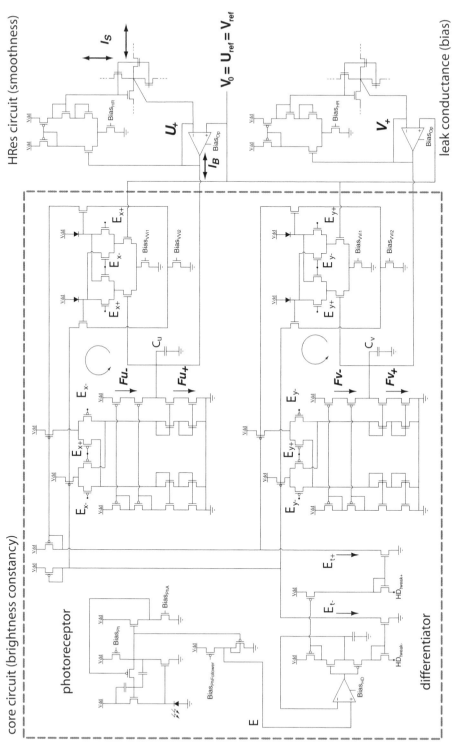

Figure 5.8 *Circuit diagram of the optical flow unit.*

circuit are ambitious: it must operate on four quadrants and should exhibit a wide linear range with respect to each multiplicand; offsets should be minimal because they directly lead to quantitative errors in the optical flow estimate; and in order to guarantee a small pixel size, the design should be compact because two of these circuits are needed per pixel. The standard Gilbert multiplier circuit [Gilbert 1968] implemented with MOSFETs meets these demands fairly well, with the exception that it only has a small linear range when operated in sub-threshold. In above-threshold operation, however, its linear range is significantly enlarged due to the transconductance change of the individual differential pair.[7] This property is used in the modified Gilbert multiplier circuit shown in Figure 5.9. The circuit embeds the original multiplier within an additional differential pair. The idea is to operate the Gilbert multiplier above threshold but use the outer, additional differential pair to rescale its output currents to sub-threshold level so that the currents in the feedback loop match.

The original Gilbert multiplier basically consists of a stack of three differential pairs. Its input is given by the differential voltages $\Delta V_A = V_1 - V_2$ applied to the lower and $\Delta V_B = V_3 - V_4$ to the two upper differential pairs. The total current I_{b1} flowing through all branches in the multiplier is controlled by the gate voltage V_{b1} of the bias transistor. Assume that I_{b1} is large enough so that all three differential pairs operate above threshold. Furthermore, assume that the differential input voltages are small enough so that all transistors are in saturation. Then, the currents flowing in the two legs of the lower differential pair are given as

$$I_1 = \frac{\beta \kappa}{8} \left(\Delta V_A + \sqrt{\frac{4 I_{b1}}{\beta \kappa} - \Delta V_A^2} \right)^2 \quad \text{and} \quad I_2 = \frac{\beta \kappa}{8} \left(-\Delta V_A + \sqrt{\frac{4 I_{b1}}{\beta \kappa} - \Delta V_A^2} \right)^2.$$

These currents are the bias currents for the upper two differential pairs. Thus, the components of the differential output current are

$$I_+ = I_{1,3} + I_{2,4} = \frac{\beta \kappa}{8} \left[\left(\Delta V_B + \sqrt{\frac{4 I_1}{\beta \kappa} - \Delta V_B^2} \right)^2 + \left(-\Delta V_B + \sqrt{\frac{4 I_2}{\beta \kappa} - \Delta V_B^2} \right)^2 \right]$$

and

$$I_- = I_{1,4} + I_{2,3} = \frac{\beta \kappa}{8} \left[\left(-\Delta V_B + \sqrt{\frac{4 I_1}{\beta \kappa} - \Delta V_B^2} \right)^2 + \left(\Delta V_B + \sqrt{\frac{4 I_2}{\beta \kappa} - \Delta V_B^2} \right)^2 \right].$$

Finally, the total output current of the multiplier core can be simplified as

$$I_{out}^{core} = I_+ - I_- = \frac{\beta \kappa}{2} \Delta V_B \left(\sqrt{\frac{4 I_1}{\beta \kappa} - \Delta V_B^2} - \sqrt{\frac{4 I_2}{\beta \kappa} - \Delta V_B^2} \right). \quad (5.15)$$

This equation does not permit a very intuitive understanding of the characteristics of the multiplier. A better understanding follows when both square roots in (5.15) are expanded

---

[7]See Appendix C.2 for a detailed analysis of the differential pair circuit.

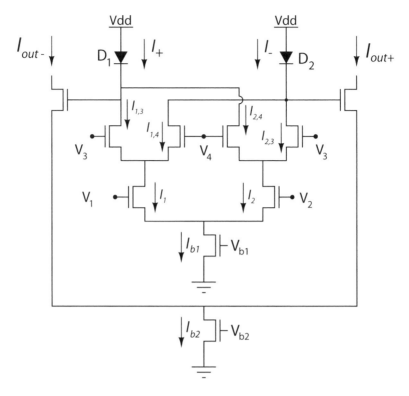

Figure 5.9 *Wide-linear-range multiplier circuit.*

The main advantage of the circuit is its compactness and the fact that its linear range and the output current level can be independently controlled. The bias current I_{b1} directly determines the linear range in above-threshold operation and is controlled with V_{b1}. Two ideal diodes transform the currents I_+ and I_- into a voltage difference that serves as the input to the outer differential pair and provides the scaled-down sub-threshold output currents I_{out+} and I_{out-}. V_{b2} controls the total output current level.

around ΔV_B^2 according to the Taylor series

$$\sqrt{a - x} \approx \sqrt{a} - \frac{x}{2\sqrt{a}} - \frac{x^2}{8a\sqrt{a}} \cdots .$$

Neglecting second- and higher order terms leads to the approximation

$$I_{out}^{core} \approx \frac{\beta \kappa}{\sqrt{2}} \Delta V_A \Delta V_B \left[1 + \frac{\Delta V_B^2}{\frac{4 I_{b1}}{\beta \kappa} - 2 \Delta V_A^2} \right]. \qquad (5.16)$$

This equation provides a better understanding of the characteristics of the Gilbert multiplier circuit. In a first approximation, the output current is proportional to the product of the two differential input voltages ΔV_A and ΔV_B, which is the goal. The term in square brackets, however, is not constant because it depends on the input. According to the input limit for the differential pair (Appendix (C.23)), the denominator has to remain positive, thus in the interval $[0, 4I_{b1}/\beta\kappa]$. The term in square brackets is always greater than one and is monotonically increasing with increasing inputs ΔV_A and ΔV_B. As a consequence, the multiplier circuit exhibits a super-linear behavior, at least as long as all transistors are in saturation and biased above threshold. The effect should be more pronounced for changes in the input voltage of the upper than the lower differential pairs because the numerator changes directly with ΔV_B^2.

Ideal diode implementation

Before the (super-)linear behavior will be examined more closely, the rest of the circuit is discussed. The differential outputs of the Gilbert multiplier circuit are connected to two diodes D_1 and D_2. Ideally, the currents I_+ and I_- each induce a voltage drop across the diodes that is logarithmically in the currents. These voltages form the input to the outer differential pair which is biased in the sub-threshold regime according to the bias voltage V_{b2}. Applying the Shockley equation

$$I_{diode} = I_0 \left[\exp\left(\frac{1}{n} V_d \frac{q}{kT} \right) - 1 \right] \tag{5.17}$$

to model the characteristics of the diodes, the output current of the entire multiplier circuit is given as

$$I_{out} = I_{out+} - I_{out-} = I_{b2} \tanh\left(\kappa n \frac{\log\left((I_+ + I_0)/(I_- + I_0)\right)}{2} \right), \tag{5.18}$$

where κ is the gate efficacy[8] of the two nFETs of the outer differential pair and $1/n$ is the ideality factor of the diodes. For arbitrary values of κ and n, I_{out} is some exponential function of the above-threshold output currents of the multiplier core as illustrated in Figure 5.10. However, if $\kappa n = 1$, Equation (5.18) simplifies to

$$I_{out} = I_{b2} \frac{I_+ - I_-}{I_+ + I_- + 2I_0}. \tag{5.19}$$

Since I_0 is very small compared to the above-threshold bias current $I_{b1} = I_+ + I_-$ it can be neglected, and so finally

$$I_{out} \approx \frac{I_{b2}}{I_{b1}} (I_+ - I_-). \tag{5.20}$$

The output of the wide-linear-range multiplier circuit is proportional to the output of the Gilbert multiplier operating above threshold, scaled by the ratio of the bias currents I_{b2}/I_{b1}.

In general, however, κ of the outer nFETs and the ideality factor $1/n$ of the diodes do not match. Because the voltage drop across the diodes in the multiplier circuit is typically within one volt, the gate–bulk potentials of the nFETs of the embedding differential pair are large. In this voltage range, κ asymptotically approaches unity because the capacitance of the depletion layer becomes negligibly small compared to the gate–oxide capacitance.

---

[8]Sometimes also referred to as the *sub-threshold slope factor* [Liu et al. 2002]. See Appendix C for more details.

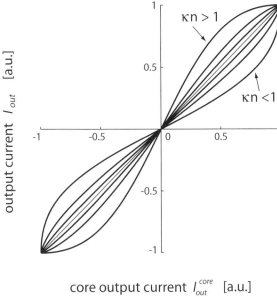

core output current I_{out}^{core} [a.u.]

Figure 5.10 *Scaling of the multiplier output current.*
The normalized output current I_{out} is displayed as a function of the core output current $I_{out}^{core} = I_+ - I_-$ for different values of $\kappa n = [0.5, 3/4, 9/10, 1, 10/9, 4/3, 2]$. The curves are computed according to (5.18) with $I_{b2}/I_{b1} = 1$ where the differential currents $I_+, I_- \in [0 \ldots 1]$, and $I_{b1} = I_+ + I_- = const$.

Thus, Equation (5.20) becomes a valid approximation if the diodes D_1 and D_2 are ideal ($n = 1$).

Ideal diodes can be implemented in different ways. A not very reasonable way is to use diode-connected pFETs, because, in order to operate them in the desired exponential domain (5.17) with $n = 1$, these pFETs must be operated sub-threshold. The diode currents, however, must be high enough so that all transistors in the multiplier core are biased above threshold. Thus, in order to remain sub-threshold, the pFETs would need to be of extreme dimensions (large W/L ratio), which is not compatible with a compact pixel design.

A compact solution is to use pn junction diodes, which are exhibiting ideal diode characteristics to a very good approximation [Gray and Meyer 1993]. While pn junctions are the basic building element in semiconductor circuits, the particular arrangement in the wide-linear-range multiplier significantly constrains possible implementations of these junctions. Specifically, only the base–emitter junction of a bipolar transistor is applicable where the collector is shorted with the base in order to prevent uncontrolled and potentially high collector current densities. Base–emitter junctions can be exploited using either native bipolar transistors in a generic BiCMOS process or vertical bipolar transistors in standard CMOS technology (see Figures 5.11a and b). Vertical bipolar transistors are intrinsically present in all well transistors as schematically shown in Figure 5.11b. The use of vertical bipolar transistors is limited because they share a common collector which is the substrate.

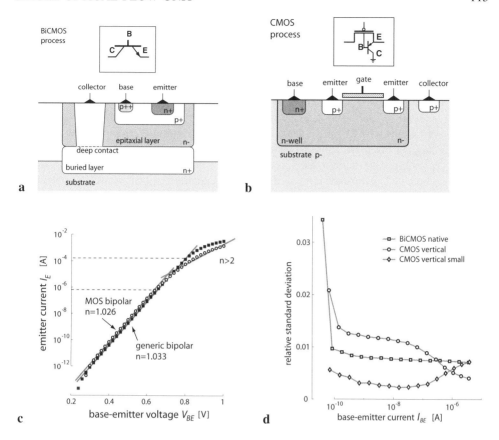

Figure 5.11 *Native bipolar versus vertical CMOS bipolar transistor.*
(a) Silicon cross-section of a native npn-bipolar transistor. Generic BiCMOS processes usually use a highly (n+) doped buried layer to achieve low collector resistivity and thus fast switching. (b) The intrinsic vertical pnp-bipolar in a pFET of a standard p-substrate CMOS process. Considering either the drain or source as the emitter, the well serves as the base and the substrate as the common collector. Obviously, a common collector constrains the use of multiple vertical bipolar transistors. (c) The measured emitter currents for the native npn- and the vertical pnp-bipolar of the pFET are shown. In both cases, base and collector are shorted and the emitter potential is moved. The deviation from the exponential characteristics for the pFET bipolar occurs at 1 μA, below the range in which the wide-linear-range multiplier is preferably biased. This deviation is mainly caused by high-level injection due to the relatively lightly doped well (base). (d) The measured relative standard deviation ($N = 6$) for the native bipolar ($A_E = 4\,\mu$m^2) and for two pFETs with two different emitter sizes $(4\,\mu$m$)^2$ and $(2.6\,\mu$m$)^2$.

As a consequence, only one type of vertical bipolar transistors is applicable depending on the substrate doping (e.g. only pnp-bipolars with a p-doped substrate), which eventually would require inversion of the entire multiplier circuit; that is, replacing all native with well transistors and switching the rails.

It is important that the bipolar implementation of the diodes D_1 and D_2 guarantees an ideal behavior up to sufficiently high current levels. Figure 5.11c shows the measured base–emitter voltages V_{BE} as a function of the applied emitter currents I_E for both a vertical pnp-bipolar transistor in a typical p-substrate CMOS process and a native npn-bipolar transistor in a generic BiCMOS process. Clearly, the two emitter currents depend exponentially on the base–emitter potential for low current values with fitted slope values very close to ideal ($n = 1.026$ and 1.033, respectively). At current levels above $1\,\mu A$, however, the vertical bipolar begins to deviate significantly from its exponential characteristics due to high-level injection caused by the relative light doping of the well (base) [Gray and Meyer 1993]. This current level is below the range in which the multiplier core is preferably operated. The exponential regime of the native bipolar, however, extends to emitter currents larger than $0.1\,mA$. Although a pure CMOS implementation of the diodes would be preferable to avoid the more complex and expensive BiCMOS process, high-level injection causes a computationally distorted behavior of the multiplier circuit. In order to guarantee an optimal implementation of the multiplier circuit a BiCMOS process is necessary. The diodes, however, are the only BiCMOS element of the whole optical flow unit.

The computational consequences of a non-ideal implementation of the diodes is rather difficult to quantify. To give some intuition of how the multiplier operation qualitatively changes, let us assume that the saturation region of the emitter currents can be approximated to first order with the Shockley Equation (5.17) with values of $n > 1$. The measured saturation of the emitter current in the vertical bipolar (Figure 5.11c) above $1\,\mu A$ is equivalent to a value $n \geq 2$. As previously shown in Figure 5.10, for increasing values $\kappa n > 1$ the scaling of the multiplier output current becomes increasingly non-linear following a sigmoidal saturation characteristic. This is unfavorable because it reduces the increase in linear range gained by the above-threshold operation of the multiplier core. In the worst case, it is imaginable that for high values of n the increase in linear range is completely counter-balanced by the non-ideal diode behavior. Thus, in order to profit fully from the increased linear range of the circuit, the implementation must provide diodes that are close to ideal at high current levels.

Output characteristics

Figure 5.12 shows the measured output characteristics of the wide-linear-range multiplier circuit for varying input voltages ΔV_A and ΔV_B respectively. At a first glance, the output current exhibits a similar sigmoidal behavior as the original Gilbert multiplier operated sub-threshold, but with a substantially increased linear range. A closer examination reveals the small predicted super-linearity in the response curves. Although the effect is almost not noticeable when varying ΔV_A, it becomes more pronounced when varying input ΔV_B as shown in Figure 5.12d. The derived analytical description for the output (5.15) nicely matches the measured curves. The measurements verify the linear scaling of the output currents to sub-threshold level according to (5.20). Of course, the behavior is only described accurately as long as all transistors in the multiplier core are saturated, and operate in the above-threshold regime.

There are lower limits for the input voltages ΔV_A and ΔV_B below which some transistors of the wide-linear-range multiplier circuit fall out of saturation. These input limits also define the range for which the previous analysis is valid. Similar considerations as for

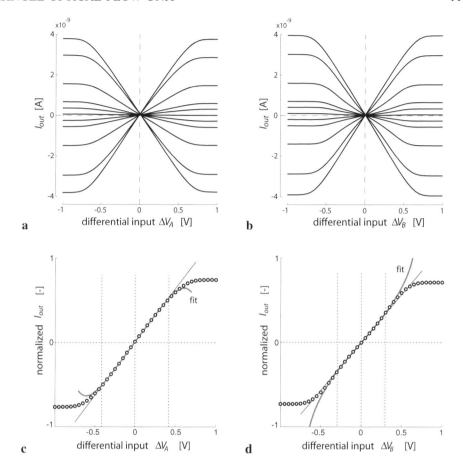

Figure 5.12 *Output characteristics of the wide-linear-range multiplier.*
(a) Measured output currents of the wide-linear-range multiplier as a function of the applied input voltage at the lower differential pair ΔV_A. (b) The same measurement, but now sweeping the input voltage ΔV_B at the two upper differential pairs. In both plots, each curve represents the output current for a fixed differential voltage ΔV (0, 0.05, 0.1, 0.15, 0.2, 0.3, 0.4, 0.5, 0.75 V) applied at the non-swept input. The bias voltages were $V_{b1} = 1.2$ V and $V_{b2} = 0.725$ V respectively. (c) and (d) Single traces for sweeping ΔV_A and ΔV_B are shown, together with a least-squares error fit according to (5.15) (bold line). The fit accurately characterizes the output currents (normalized to total bias current I_{b2}) for input voltages within the input limits (vertical dotted lines).

the simple differential pair circuit (Appendix C) lead to the following *saturation conditions* for the multiplier circuit:

$$\max(V_1, V_2) > V_{b1} + (V_{b1} - V_{T0}) \qquad \text{and}$$

$$\max(V_3, V_4) > \max(V_1, V_2) + (V_{b1} - V_{T0}). \tag{5.21}$$

These are important operational limits that the design of the single optical flow unit has to account for. For example, the additional unity-gain source follower at the photoreceptor output permits the shift of the output voltage to a sufficiently high DC level (see Figure 5.8). Because the photoreceptor output voltages from nearest-neighbor optical flow units form the input to the upper differential pairs (V_3, V_4) of the multiplier circuits, these voltages have to obey (5.21). The input to the lower differential pair is given by the differential voltage $U_+ - V_0$ and $V_+ - V_0$ respectively. Since the virtual ground V_0 is adjustable, it can be set so that it meets conditions (5.21).

In order to keep all transistors of the multiplier circuit core in above-threshold operation, the maximal differential input voltages are limited to

$$\Delta V_A < (V_{b1} - V_{T0}) \quad \text{and} \quad \Delta V_B < \frac{\sqrt{2}}{2}(V_{b1} - V_{T0}). \quad (5.22)$$

These *input limits* (represented by the dotted vertical lines in Figures 5.12c and d) correspond well with the measured voltage range within which Equation (5.15) accurately describes the output currents.

As Figure 5.12 illustrates, the linear output range of the multiplier circuit is almost an order of magnitude larger than the one of the standard Gilbert multiplier running sub-threshold (≈ 100 mV). Figure 5.13a illustrates how the linear range depends on the bias current I_{b1}. Here, the linear range is defined as the input voltages for which the output current deviates by less than 5% from a linear fit through the origin. As shown in Figure 5.13a, the linear range for input ΔV_B is substantially smaller than for input ΔV_A. The reason is the pronounced super-linear behavior of the multiplier output for input ΔV_B (5.16). In the particular example shown in Figure 5.13b, it is strong enough to exceed the applied linearity measure and thus leads to the low values. The more important deviation from linearity, however, is saturation at large inputs. Saturation occurs when the inputs ΔV_A, ΔV_B exceed the input limits (5.22). Defined again as the voltage level at which the output deviates by more than 5% from a linear fit through the origin (but neglecting the superlinear behavior), the measured input range for which the multiplier is not saturating is almost equally large for both inputs. It is slightly beyond the linear range found for ΔV_A as shown in Figure 5.13a. The output saturation defines the major computational limitation of the wide-linear-range multiplier circuit.

The increase in linear range as a function of bias current can be modeled. In particular, the maximal slope of the normalized output currents for given inputs ΔV_A, ΔV_B is proportional to the transconductance of the multiplier circuit with respect to each of the two inputs. Differentiating the core output current (5.15) with respect to ΔV_A and ΔV_B, respectively, leads to

$$g_{mA} = \frac{\partial I_{out}}{\partial \Delta V_A}\bigg|_{\Delta V_A \to 0, V_B = const.} = \frac{1}{\sqrt{2}} \frac{I_{b2}}{I_{b1}} \beta \kappa \Delta V_B \sqrt{\frac{I_{b1}}{I_{b1} - \beta \kappa \Delta V_B^2 / 2}} \quad (5.23)$$

and

$$g_{mB} = \frac{\partial I_{out}}{\partial \Delta V_B}\bigg|_{\Delta V_B \to 0, V_A = const.} = \frac{1}{\sqrt{2}} \frac{I_{b2}}{I_{b1}} \beta \kappa \Delta V_A. \quad (5.24)$$

With the assumption that the output currents represent similar sigmoidal functions, the linear range for each of the two input voltages is inversely proportional to the associated

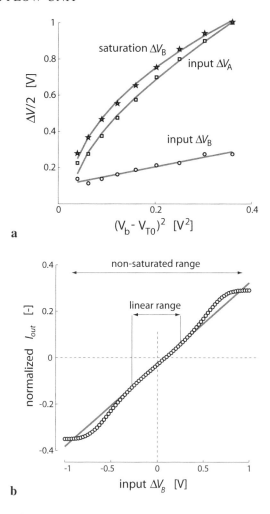

Figure 5.13 *Linear and non-saturating output range.*
(a) The measured linear output range of the wide-linear-range multiplier circuit is plotted as a function of the bias current I_{b1} for both input stages. Linearity is massively increased in comparison to the sub-threshold regime ($\approx$100 mV). The linear range grows proportionally to the inverse of the transconductance of the multiplier circuit, illustrated by the least-squares error fit based on (5.23) and (5.24). Due to the super-linear behavior, the measured linear range (5% deviation) is smaller for variable input ΔV_B. In practice, however, saturation is of much greater importance because it limits the potential currents in the feedback loop.
(b) The output current as a function of input ΔV_B clearly shows the super-linear behavior that is pronounced at high bias currents ($V_{b1} = 1.4$ V).

transconductance. The two fitting functions, based on the inverse of the above transconductances (5.23) and (5.24), describe fairly well the characteristics of the measured linear ranges as shown in Figure 5.13a.

Saturation in the feedback loop

The output current of the multiplier circuit saturates for large input voltages. What does this mean in terms of the expected motion output of a single optical flow unit?

For the sake of illustration, let us assume once more a 1-D array of the optical flow network. Furthermore, all lateral connections in this network are disabled and the bias constraint neglected ($\rho = 0, \sigma = 0$). Then, for given spatio-temporal brightness gradients, the circuit ideally satisfies the brightness constancy constraint

$$E_x u + E_t = 0. \tag{5.25}$$

Now, the multiplication $E_x u$ is replaced by the saturating function $f(u)|_{E_x}$ with E_x constant, which describes the output characteristics of the wide-linear-range multiplier circuit.[9] For reasons of simplicity and because this discussion is of a qualitative nature, assume that the output I_{out}^{core} of the multiplier core follows a sigmoidal function (*e.g.* tanh(u)) for any given E_x. The multiplier output (5.20) can now be rewritten as $f(u)|_{E_x} = c_{E_x} I_{b2}/I_{b1} \tanh(u)$ where the constant c_{E_x} is proportional to a given E_x. Since $f(u)$ is one-to-one, (5.25) can be solved for the motion response, and hence

$$u = f^{-1}(-E_t) = -\text{artanh}\left(\frac{I_{b1}}{I_{b2}} E_t c_{E_x}^{-1}\right). \tag{5.26}$$

Figure 5.14 illustrates the expected motion output according to (5.26) for different values of the bias current I_{b2}. The motion output increasingly over-estimates the stimulus velocity the more the multiplier circuit starts to saturate until – in a real circuit – it finally hits the voltage rails. The more the output of the multiplier circuit is below the true multiplication result, the more the negative feedback loop over-estimates u in order to assert the brightness constancy constraint (5.25).

With respect to the true above-threshold characteristics of the real multiplier circuit, the responses are expected to differ from the curves in Figure 5.14 only insofar as the linear range is increased and the saturation transition is sharper and more pronounced. Increasing I_{b2} decreases the slope of the motion response. Thus, the bias current of the outer differential pair acts as a gain control that permits, for example, adjustments of the linear range of the motion response to the expected speed range of visual motion, so obtaining the highest possible signal-to-noise ratio. This adjustment, however, remains a manual operation.

The wide-linear-range multiplier circuit provides compact four-quadrant multiplication where the linear/saturation range and the multiplication gain can be separately controlled by two bias voltages V_{b1} and V_{b2}. The disadvantages are the increased power consumption caused by the above-threshold operation of the core, and the need for BiCMOS process technology to implement diodes that show an ideal behavior up to sufficiently high current densities.

---

[9]According to the notation in the schematics of the optical flow unit (Figure 5.8), the inputs to the multiplier circuit are the differential voltages $E_x = E_{i+1} - E_{i-1}$ (voltage difference of nearest-neighbor photoreceptors) and $u = U_+ - U_0$.

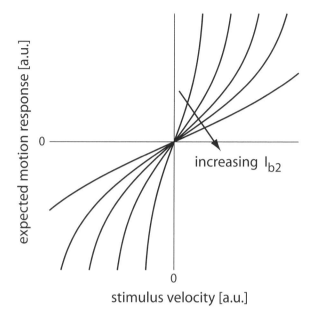

Figure 5.14 *Expected motion output due to saturation in the multiplier circuits.*
The figure qualitatively shows the expected speed tuning curves according to
Equation (5.26). Note that the more the multiplier saturates, the more the motion out-
put expands. The different curves correspond to different values of the bias current I_{b2}.
The bias voltage V_{b2} controls the response gain and thus permits the linear output range to
be adjusted to the expected velocity range.

5.4.2 Effective bias conductance

The quadratic form of the bias constraint in the cost function (4.19) results in an ohmic bias
constraint current that is proportional to the constant conductance σ. Sufficiently low and
compact ohmic conductances, however, are difficult to implement in CMOS technology. It is
much easier and more convenient to use small transconductance circuits to implement these
conductances. As shown in the circuit diagram (Figure 5.8), the bias conductance of the
optical flow unit is implemented as the transconductance of a differential transconductance
amplifier circuit configured as a unity-gain follower. Instead of an ohmic current (5.14),
it generates a saturating bias current $I_B(\Delta V)$ that is a function of the voltage difference
between the voltages U_+ and V_+ respectively, and V_{ref}, the voltage that reflects the reference
motion. The current characteristics are ideally

$$I(\Delta V) = I_b \tanh\left(\frac{\kappa}{2U_T}\Delta V\right), \tag{5.27}$$

with I_b being the bias current of the amplifier circuit controlled by the voltage Bias$_{OP}$, U_T
the thermal voltage, and κ the common sub-threshold slope factor. The transconductance
$g_m = \frac{I_b\kappa}{2U_T}$ characterizes the small ohmic regime and I_b determines the saturation current

level of the circuit. The sigmoidal current characteristics change the bias constraint continuously from a quadratic to an absolute-value function for increasing voltage differences, which has computational advantages as discussed in detail in Section 4.4.

Unfortunately, the follower-connected amplifier does not exclusively constrain the bias conductances. Consider the current equilibrium at the capacitive nodes C_u, C_v in the single optical flow unit. At steady state, the currents onto these nodes represent the different constraints according to the dynamics (4.25). To be completely true, this would require the feedback current to be generated by ideal current sources. The small but present output conductances of the feedback loop, however, lead to a deviation of such an ideal model. Figure 5.15b shows one of the two feedback nodes of the single motion unit. An incremental increase of V requires the total feedback current $F_+ - F_-$ to be larger than actually defined by the constraint dynamics, by an amount that is proportional to the sum of the output conductances g_{oN} and g_{oP}. The extra current has to be provided by the feedback loop and thus alters the computational result. Since this current depends proportionally on the voltage V, it acts like a *second* bias constraint current that is added to the bias current I_B of the transconductance amplifier. The reference potential of this second bias conductance is intrinsic and cannot be controlled. It depends on various parameters like the total current level in the feedback loops $(F_+ + F_-)$ or the Early voltages of the transistors. In general, this reference voltage is not identical with V_{ref}. Thus, the superposition of the two bias constraint currents has an asymmetric influence on the final motion estimate. Since the total feedback currents are typically weak, the effect is significant. The aim is to reduce the output conductances as much as possible, that is to reduce the output conductances g_{oN} and g_{oP} as much as possible.

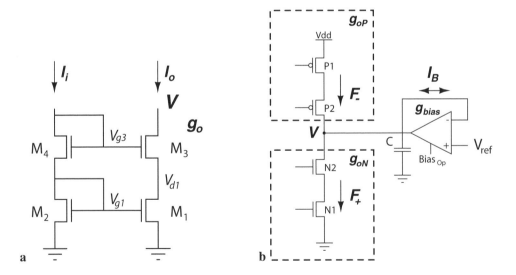

Figure 5.15 *Cascode current mirror and the effective bias conductance.*
(a) An n-type cascode current mirror. (b) The effective bias conductance is the combination of the output conductances of the current mirrors g_{oN} and g_{oP} and the transconductance of the operational amplifier g_{bias}.

A cascode current mirror circuit has a significantly lower output conductance than a simple current mirror. Figure 5.15a shows a single native-type cascode current mirror. Using the labels from the figure and neglecting the junction leakage in drain and source, the output conductance can be written as

$$g_o = g_{o1} \frac{g_{ds3}}{g_{ms3}} \qquad (5.28)$$

where g_{ds3} and g_{ms3} are the drain and transconductance of transistor M_3 and g_{o1} is the output conductance of transistor M_1. The decrease in output conductance compared to g_{o1} (the output conductance of a simple current mirror circuit) is given by the ratio

$$\frac{g_{ds3}}{g_{ms3}} = \frac{V_{E,M3}}{U_T} \approx 250 \ldots 1000 \qquad (5.29)$$

where U_T is the thermal voltage kT/q and $V_{E,X}$ denotes the Early voltage of transistor X. Applying the same analysis to the inverted current mirror, the final output conductance (see Figure 5.15b) becomes

$$g_o = g_{oN} + g_{oP} = \frac{F_+ \cdot U_T}{V_{E,N1} \, V_{E,N2}} + \frac{F_- \cdot U_T}{V_{E,P1} \, V_{E,P2}}. \qquad (5.30)$$

The output conductance depends equally strongly on both transistors in the cascode current mirror. Since the Early voltage is proportional to the length of a transistor[10] both transistors should be long in order to reduce the output conductance. If restrictions in implementation area do not allow this, then, because of matching reasons, the length of the mirror transistor (M_1, M_2) preferably has to be increased at the expense of the length of the cascode transistor. The total output conductance depends also on the strength of the feedback currents F_+ and F_-. These currents are mainly determined by the bias current of the multiplier circuit. A high bias voltage $Bias_{VVI2}$ therefore amplifiers the unwanted second-order effects.

Also, the current induced by the output conductance (5.30) does not saturate, in contrast to the current I_B. Thus, even when the bias conductance g_{bias} is determining the behavior around V_{ref}, it can be that the output conductance dominates at higher motion signals. This can lead to asymmetric responses at high speeds and high feedback current densities as seen in the next chapter (Figure 6.4b). It is important to minimize the output conductances of the feedback loop as much as possible.

5.4.3 Implementation of the smoothness constraint

The smoothness conductance ρ is implemented as the so-called horizontal resistor (HRes) circuit [Mead 1989]. The current–voltage characteristic of this element is basically equivalent to the one of the follower connected transconductance amplifier circuit. The non-linear implementation of the smoothness conductance has advantages, as discussed in Section 4.4. The HRes circuit reduces smoothing across large voltage differences. The smoothness constraint changes from a quadratic to an absolute-value function as the voltage difference enters the saturation region. As shown in the schematics in Figure 5.8, the HRes circuit

---

[10]See Appendix C.

consists of a single transconductance amplifier with an additional diode-connected transistor in the output leg and four pass transistors, which implement the resistive connections to the four neighboring optical flow units.

Another useful feature of the HRes circuit is the fact that the conductance strength is adjustable. For voltage differences within the small-signal range of the amplifier, the conductance is proportional to the bias current controlled by the gate voltage $Bias_{HR}$. Above the linear range, the current through the pass transistor saturates and is equal to half of the bias current. Although the bias voltage $Bias_{HR}$ sets the saturation current globally, that is for all lateral connections in the optical flow network, the smoothing is anisotropic and locally controlled by the voltage differences (optical flow gradients) in the resistive layers of the optical flow network.

5.5 Layout

The last step in the design of a silicon implementation of the optical flow network is creating a layout for circuit production. The layout determines in great detail the geometry of all circuits and connections on-chip. The layout specifies exactly how to generate the different processing masks that determine, for example, where and in which size a transistor gate is deposited, or which areas of the substrate are doped with acceptors or donors.

Unlike implementations of digital circuits, generating the layout for analog sub-threshold circuits is typically not an automated process. With the exception of a few simple and standardized building blocks in the chip periphery, each individual transistor and connection of the chip layout is specifically designed by the analog designer. The exact dimensions of the transistors are crucial for the computational behavior of the analog circuits. Also the shape and spatial arrangement are important. Imperfections in the fabrication process usually lead to mismatch, which describes the fact that identically designed transistors can be effectively slightly different in size after fabrication, thus having different drain currents for the same terminal voltages. Also doping irregularities (gradients) can alter transistor properties such that drain current densities vary across the silicon area. And last but not least, the very small sub-threshold current signals are susceptible to noise of any kind. A good and finally successful layout is considerate of all these imperfections and minimizes their impact on the computational properties of the circuit. Creating a good layout usually takes time and experience, but obviously depends also on the particular circuit. Some standard tricks and guidelines can be found in the literature (see *e.g.* [Liu et al. 2002]).

Figure 5.16 shows the layout of a single optical flow unit. It basically has to account for two conflicting constraints: the layout should be as compact as possible yet show maximal functional quality. Compactness is important because a compact pixel design increases the resolution of the whole optical flow network per silicon area. Also, if the photodiode size is kept constant, a more compact layout is equivalent to a higher fill-factor. However, compactness opposes the design of large transistors, which would reduce mismatch. A good design minimizes the dimensions of those transistors for which matching is less important. Naturally, the layout has to obey the particular specifications of the fabrication process used. Individual processes differ in the number of available layers (such as metal layers) and also have specific design rules that the layout has to obey.

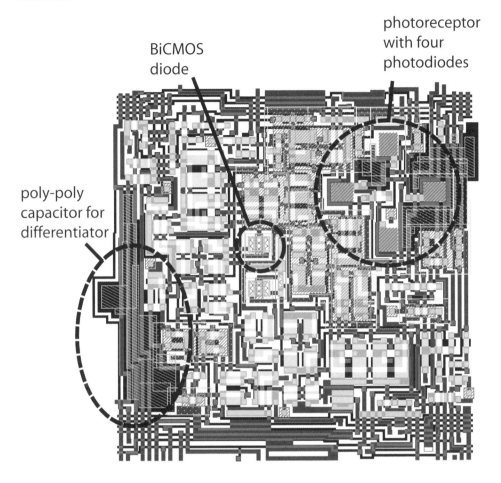

Figure 5.16 *Layout of a single optical flow unit.*
In order to improve readability, not all layout layers are shown. Some particular circuit devices are specifically labeled. Note that a substantial amount of silicon real-estate is occupied by wires. More detailed illustrations can be found on the http://wiley.com/go/analog.

Although the different layers in Figure 5.16[11] are difficult to distinguish on the gray-scale image, and some are left out for reasons of readability, it is sufficient to point out some particularities of the layout. The photodiode, located in the upper right corner, is split into four equally sized rectangular areas, forming a cross-like arrangement. That way, the visual information is averaged over a larger area. The photodiodes were the only areas that a top-level metal layer did not cover. This metal layer shields the rest of the circuitry from impinging light that can cause unwanted photocurrents.

Another design characteristic of the layout is its approximate symmetry along the (lower left to upper right) diagonal. Since the optical flow unit consists basically of twice the same

---

[11] Detailed process specifications are given in Appendix D.

circuitry for each of the two optical flow components, the layout is kept as symmetric as possible to ensure best possible matching. The temporal differentiator circuit and the photoreceptor, the only two unique sub-circuits, are therefore placed at opposite corners. Transistors for which matching is important are increased in size, for example the transistors of the wide-linear-range multipliers or the cascode current mirrors.

And finally, the layout in Figure 5.16 clearly illustrates that wires are occupying a significant amount of the total silicon real-estate needed. The nearest-neighbor connections of the two-layer resistive network, the spatial gradient estimation as the difference operator of the nearest-neighbor photoreceptor signals, and the substantial number of bias voltage and read-out lines lead to crowded conditions. It is obvious that a network design with denser and more far-reaching connections would lead to an even higher fraction of silicon real-estate spent only on wires. Wires are costly!

6

Smooth Optical Flow Chip

The smooth optical flow chip is the silicon realization of the optical flow network. It contains a homogeneous array of optical flow units. Because of the modular layout, the array size and thus the resolution of the optical flow estimate can be almost arbitrary. This chapter carefully characterizes the smooth optical flow chip and shows how it behaves under real-world visual conditions. Two versions of the smooth optical flow chip have been produced so far which are essentially identical in the layout of the single optical flow unit. They only differ in their array sizes where the smaller chip has 10×10 and the larger 30×30 units, respectively. These are the effective array sizes, that is the number of functional optical flow units. On each side of the array, there is an additional row of photoreceptor circuits necessary for computing the spatial brightness gradients of the edge units. Exact specifications of the two circuits and the process technology are listed in Appendix D.

In addition to the optical flow array, the chip also includes peripheral circuitry for power supply and biasing of the circuits, and to read out the signals of the optical flow network. There is no possibility to have a simultaneous read-out of all units in the optical flow array because that would require an unfeasible number of wires. Instead, a scanner circuit permits sequentially addressing of each unit such that they can be read out one at a time [Mead and Delbrück 1991]. As outlined in Figure 6.1, the scanner consists of a horizontal and a vertical shift register located at two edges of the array. According to an externally applied clock signal, the scanner addresses a different unit at each clock cycle and reroutes voltage follower circuits to read out the local photoreceptor and the optical flow signal. Read-out rates up to 5000 frames per second are possible with the current design.

A direct output connection to one of the corner units of the optical flow array permits the continuous read-out of the global optical flow estimate – if the smoothness conductances are set high. This is an interesting way to interface the smooth optical flow chip for applications where an external clock signal is too costly to generate and/or a global optical flow estimate is sufficient.

Analog VLSI Circuits for the Perception of Visual Motion A. A. Stocker
© 2006 John Wiley & Sons, Ltd

Figure 6.1 *Read-out and peripheral circuitry of the smooth optical flow chip.*
The read-out circuitry of the smooth optical flow chip consists of a horizontal voltage scanner (and a vertical shift register) that permits the sequential read-out of the local optical flow estimate and the photoreceptor signal of each unit. The remaining peripheral circuitry includes the pad circuits for power supply and bias voltages. The chip also permits a direct read-out of the lower left corner unit of the optical flow network. In global mode, *i.e.* when the smoothness conductances are set very high, this provides the global optical flow estimate without using the scanner circuits.(Analog Integrated Circuits and Signal Processing, Volume 46, No. 2, 2006, Analog integrated 2-D optical flow sensor, Stocker, A. Alan, with kind permission of Springer Science and Business Media.)

6.1   Response Characteristics

The smooth optical flow chip has been carefully characterized. The chip was mounted on an optical test-bench in a dark-room environment that allowed complete control over stimulus conditions. For all measurements the stimuli were projected directly onto the chip via an optical lens system. The stimuli either were generated electronically and displayed on a computer screen, or were physically moving objects, for example printed patterns on a rotating drum. The adaptive properties of the photoreceptor circuit make the smooth optical flow chip relatively insensitive to the absolute irradiance level. The measured on-chip irradiance in all the experiments was within one order of magnitude with a minimal value of 9 mW/m^2 for very low-contrast computer screen displays.[1] At such low values, the rise time of the photoreceptor circuit is in the order of a few milliseconds [Delbrück 1993b]. This is sufficiently fast considering the applied stimulus speeds in those particular measurements. The bias voltages are named according to the labels provided in the circuit diagram shown in Figure 5.8. Unless indicated differently, the bias voltages were all sub-threshold and were kept constant at their mentioned default values. The measured optical flow output signals are always indicated with respect to the virtual ground V_0.

Characterization of the single optical flow unit has been performed using the chip with the smaller array. Moving sinewave and squarewave gratings were applied to characterize the motion response for varying speed, contrast, and spatial frequency of the gratings. Furthermore, the orientation tuning of the smooth optical flow chip and its ability to report the intersection-of-constraints estimate of image motion were tested. In order to increase the robustness of these measurements, the presented data constitute the global motion signal, thus the unique, collectively computed solution provided by the 10×10 units of the smooth optical flow chip. Data points represent the average of 20 measurements, where each measurement constitutes the mean signal over one cycle of the grating stimuli. Consequently, measured standard deviations are rather small and omitted in the plots where they were not informative.

6.1.1   Speed tuning

The speed tuning of the smooth optical flow unit was tested using drifting sine- and squarewave gratings of high contrast (80%) and low spatial frequency (0.08 cycles/pixel). Figure 6.2 shows the chip's motion output as a function of stimulus speed. Negative values indicate stimulus motion in the opposite direction. Only the component of the optical flow vector that is orthogonal to the grating orientation is displayed. The data confirm the prediction that saturation in the multiplier circuit leads to an expansive non-linear motion estimate (compare to Figure 5.14). The response is approximately linear in stimulus speed within the voltage range compared to the non-saturated operation range of the multiplier circuit. For the given bias strength of the multiplier core ($Bias_{VVI1} = 1.1$ V), the linear range is approximately ± 0.5 V as indicated by the dashed lines. Beyond this linear range, the output quickly increases/decreases and finally hits the voltage rails in either direction (not shown). Within a small range around zero motion, the optical flow output slightly underestimates the stimulus velocity. The differences in response to a sine- and a squarewave grating are minimal.

The next two graphs in Figure 6.3 show speed tuning for the same sine- and squarewave stimuli, but now measured for different settings of the bias voltage $Bias_{VVI2}$. As predicted from the analysis in Chapter 5, increasing the multiplier bias current leads to a smaller response gain, thus expanding the speed range for which the response is linear. The solid lines represent the linear fits to the individual measurements. The slope of these lines roughly scales by a factor of one-half for a fixed increase in bias voltage by 30 mV, thus the response gain is inversely proportional to the bias current. The bias voltage $Bias_{VVI2}$ allows the optimal adjustment of the linear operational range to the speed of the expected visual motion.

---

[1] For comparison, irradiance under bright sunlight is in the order of 120 W/m^2.

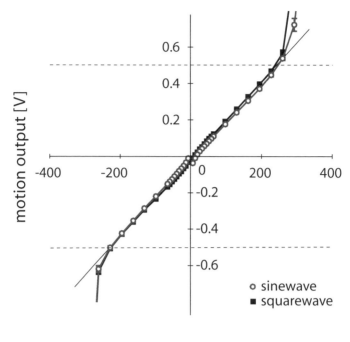

stimulus speed [pixel/s]

Figure 6.2 *Speed tuning of the smooth optical flow chip.*
The speed tuning of the smooth optical flow chip to a moving sine- and squarewave grating is shown (spatial frequency 0.08 cycles/pixel, contrast 80%). The tuning curve follows the predicted response qualitatively, showing the expanding non-linearity due to the saturation of the multiplier circuits.

It also permits the recalibration of the smooth optical flow chip if scale changes occur in the optical pathway. The slope values for the two different stimuli are almost identical (see the insets in Figure 6.3). This suggests that the speed tuning is not dependent on the spatial frequency spectrum of the stimulus as long as the fundamental frequency is the same.

Figure 6.4 reveals a closer look at both the low and high limits of correct speed estimation. For appropriate values of Bias_{VVI2}, the range of reliable speed estimation spans three orders of magnitude, from 1 pixel/s up to at least 3500 pixels/s. It is likely that the chip is able to measure even higher visual speeds, although it was not possible to verify this because the applied test setup could not produce such high visual speeds. In principle, the upper bound of accurate speed estimation is given by the low-pass filter cut-off frequency of the photoreceptor and/or the temporal differentiator. The pronounced asymmetry between the response to high speeds of opposite directions is likely due to the high current level in the feedback loop (high Bias_{VVI2}) which increases the undesired output conductances at the feedback node.

The response for very low speeds is slightly confound by offsets (Figure 6.4a). Given the low output currents from the wide-linear-range multiplier (bias voltage $\text{Bias}_{VVI2} = 0.35$ V)

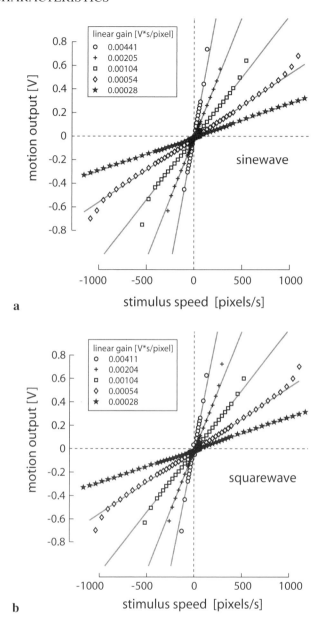

Figure 6.3 *Response gain.*
(a) The output of the smooth optical flow chip to a drifting sinewave grating is shown as a function of the multiplier bias voltage (Bias$_{VVI2}$ = 0.5, 0.53, 0.56, 0.59, 0.62 V). The response gain is inversely proportional in bias currents as indicated by the fitted slope values (exponentially in bias voltage; see insets). (b) The same measurements for a squarewave grating of the same spatial frequency and contrast. The fitted slope values are almost identical. (Analog Integrated Circuits and Signal Processing, Volume 46, No. 2, 2006, Analog integrated 2-D optical flow sensor, Stocker, A. Alan, with kind permission of Springer Science and Business Media.)

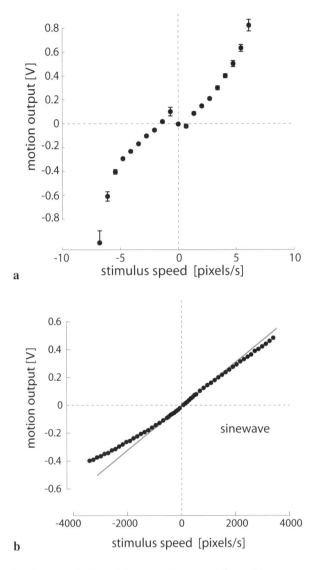

Figure 6.4 *Speed estimation limits of the smooth optical flow chip.*
(a) Speed tuning of the smooth optical flow chip when biased to detect very low speeds
(Bias$_{VVI2}$ = 0.35 V). The chip is able to resolve low speeds down to 1 pixel/s. (b) Appropriately biased (Bias$_{VVI2}$ = 0.67 V) high speeds up to 3500 pixels/s can be measured reliably.
Note the asymmetries due to the high output conductances of the feedback loop (see also
Section 5.4.2). (Analog Integrated Circuits and Signal Processing, Volume 46, No. 2, 2006,
Analog integrated 2-D optical flow sensor, Stocker, A. Alan, with kind permission of
Springer Science and Business Media.)

and considering the low currents from the temporal differentiator at such lowstimulus speeds, the total differential currents at the capacitive nodes are also low. Dynamic second-order effects dominate. However, zero motion is reliably detected since the input is constant and the bias constraint ($\text{Bias}_{OP} = 0.30$ V) pulls the output toward the reference motion. The deviations at low velocities are less pronounced at higher current densities, that is higher values of Bias_{OP} and Bias_{VVI2}.

6.1.2   Contrast dependence

The output of the smooth optical flow chip depends continuously on stimulus contrast. The lower the contrast, the lower the feedback currents, and, thus, the more dominant the influence of the bias constraint, forcing the response toward the reference voltage V_{ref} (see also Figure 4.6). Figure 6.5a shows the output voltage as a function of stimulus contrast for a drifting sinewave and squarewave grating stimulus of constant speed (30 pixels/s) and identical spatial frequency (0.08 cycles/pixel).

Below about 35% contrast, the optical flow signal clearly depends on contrast and decreases toward zero motion as contrast approaches zero. The response to squarewave gratings shows a slightly higher resistance against the fall-off. The bold line is a least-squares error fit to the sinewave grating data according to Equation (5.10). The fit does not capture the circuit's behavior exactly: the measured response curve rises faster and exhibits a flatter plateau than the fit. This has been expected because (5.10) assumes an ohmic conductance model whereas the implementation uses a non-ohmic bias conductance σ which decreases with increasing motion output signal (transconductance amplified). The effect is that the output is less affected by contrast. The fit was chosen deliberately according to Equation (5.10) to reveal the positive effect of the non-ohmic bias conductance.

The influence of the strength of the bias conductance on contrast tuning is shown in Figure 6.5b. Solid lines again represent the least-squares error fits according to Equation (5.10). Increasing values of the bias voltage Bias_{OP} strongly increase contrast dependence, even at high contrast values. For high bias voltages, the bias conductance dominates: the denominator in Equation (5.10) becomes approximately constant and reduces the computation basically to a feedforward multiplication of the spatial with the temporal brightness gradients. The motion output becomes quadratically dependent on contrast. This quadratic dependence can be observed in Figure 6.5b, but only at very low contrast levels because the simple Gilbert multipliers responsible for the product of the spatial and temporal gradient saturate early ($>100\,\text{mV}$), leading to a linear dependence on contrast. A simple spatio-temporal gradient multiplication makes the costly feedback architecture of the circuit obsolete and can be implemented in a much more compact way [Moore and Koch 1991, Horiuchi et al. 1994, Deutschmann and Koch 1998a].

6.1.3   Spatial frequency tuning

The third parameter was spatial frequency, tested for sinewave and squarewave gratings (80% contrast, 30 pixels/s). Figure 6.6a shows the chip's response as a function of increasing fundamental spatial frequency. The spatial frequency k is given in units of the Nyquist frequency $[\pi/d]$. As with contrast the difference in waveforms does not have a significant effect on the response of the smooth optical flow chip. Again, a least-squares error fit according to Equation (5.10) was performed on the sinewave grating data for frequencies

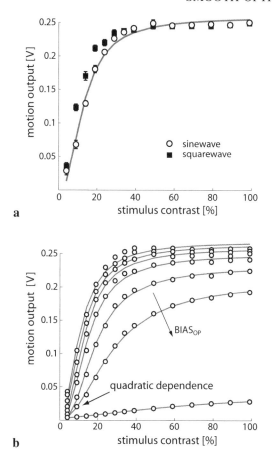

Figure 6.5 *Contrast dependence of the motion output.*
(a) The motion output shown as a function of stimulus contrast for sine- and squarewave gratings of constant speed (30 pixels/s) and spatial frequency (0.08 cycles/s). The bias constraint forces the responses toward zero motion ($V_{\text{ref}} = V_0$) for low stimulus contrast. Especially for squarewave gratings, the measured response curves exhibit a sharper onset and a flatter plateau behavior than predicted for an ohmic bias conductance. (b) Increasing values of the bias voltage Bias$_{OP}$ (0.15, 0.23, 0.26, 0.29, 0.32, 0.35, 0.45 V) lead to a stronger contrast dependence even at high contrast values, here shown for sinewave gratings only. For highest values (bottom two curves), the output signals deteriorate and become proportional to the product of the spatial and temporal brightness gradients. Due to the early saturation of the simple Gilbert multiplier, the expected quadratic dependence is only noticeable at small signal levels (<100 mV). (Analog Integrated Circuits and Signal Processing, Volume 46, No. 2, 2006, Analog integrated 2-D optical flow sensor, Stocker, A. Alan, with kind permission of Springer Science and Business Media.)

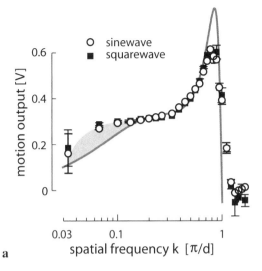

a

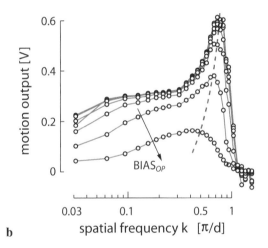

b

Figure 6.6 *Spatial frequency tuning of the optical flow chip.*
(a) The spatial frequency response was measured for sinewave and squarewave gratings of constant speed (30 pixels/s) and contrast (80%). For frequencies $k < 1$ $[\pi/d]$, the response closely follows the expected behavior (fit according to (5.10) – solid curve), except at low frequencies where the non-ohmic bias conductance of the implementation leads to an extension of the correct estimation range (shaded area). (b) Increasing the bias voltage Bias_{OP} (0.1, 0.15, 0.2, 0.25, 0.30, 0.35, 0.40 V) reduces the overall response and shifts the peak response toward $k = 0.5$ $[\pi/d]$ where the spatial gradient and thus the local contrast is highest. Only data for sinewave gratings are shown. (Analog Integrated Circuits and Signal Processing, Volume 46, No. 2, 2006, Analog integrated 2-D optical flow sensor, Stocker, A. Alan, with kind permission of Springer Science and Business Media.)

$k < 0.75$. For spatial frequencies $k < 0.06$, the local spatial gradients on the chip surface become very low, and thus the bias constraint dominates and forces zero motion (V_{ref}). Once more, the non-ohmic bias conductance improves the optical flow estimate. The output shows an increased resistance to a response fall-off in the frequency range $0.06 < k < 0.3$ compared to the fit that assumes an ohmic bias conductance. The difference (shaded area) is significant and particularly important because this low-frequency range is usually dominant in natural visual scenes. As spatial frequencies become high, $k > 1$, the response rapidly drops toward zero and remains small.

The strength of the bias voltage Bias$_{OP}$ affects the amplitude of the motion response as well as the spatial frequency for which the motion response is maximal. This is shown in Figure 6.6b. Increasing Bias$_{OP}$ decreases the motion output the more, the smaller is the local spatial gradient. Because E_x is largest at $k = 0.5$ (see also Figure 5.6) the spatial frequency for which the motion output is maximal moves progressively toward this value with increasing values of Bias$_{OP}$.

6.1.4 Orientation tuning

The ability of the smooth optical flow chip to accurately estimate two-dimensional visual motion is superior to other reported hardware implementations. Figure 6.7 shows the applied plaid stimulus and the measured global motion components U and V of the smooth optical flow chip as a function of the motion direction of the plaid. The output fits well the theoretically expected sine and cosine functions. Note that the least-squares error fit (solid lines) shows a slight total orientation offset of about 6 degrees, which is due to the imperfect calibration of the stimulus orientation with the orientation of the optical flow array.

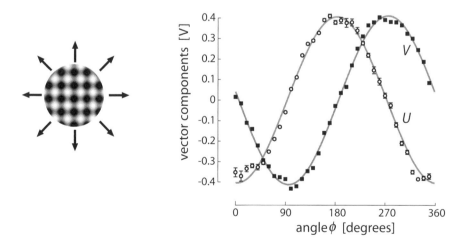

Figure 6.7 *Orientation tuning of the smooth optical flow chip.*
The smooth optical flow chip exhibits the expected cosine- and sinewave tuning for the components U and V of the global optical flow vector. The sinusoidal plaid stimulus (80% contrast, 0.08 cycles/pixel) was moving at a velocity of 30 pixels/s.

6.2 Intersection-of-constraints Solution

The real strength of the smooth optical flow chip and the feature that distinguishes it from other aVLSI visual motion sensors is its collective computational approach. It is able to find the intersection-of-constraints (IOC) solution, which is necessary for solving the aperture problem for particular stimuli. To demonstrate this, the chip was presented with a high-contrast visual stimulus that consisted of a dark triangle-shaped object on a light non-textured background moving in different directions. Because of the particular object shape, there is no other way to determine the correct image motion besides computing the IOC solution.

Figure 6.8 shows the applied stimulus and the global motion output of the chip for a given constant positive or negative object motion in the two orthogonal directions. Each data point represents the time-averaged global motion vector for a given strength of the bias constraint. For small bias conductances (Bias$_{OP}$ ≤ 0.25 V) the smooth optical flow chip reports almost perfectly the correct object motion in either of the four tested directions. A small tilt of the motion vector in the order of less than 10 degrees remains as the result of the remaining intrinsic output conductance at the feedback node (see Section 5.4.2). For comparison, the direction of the vector average of a normal flow estimate is also indicated, which deviates by 22.5 degrees from the correct object motion for this particular object shape. As the bias constraint is increased, the direction of the global flow

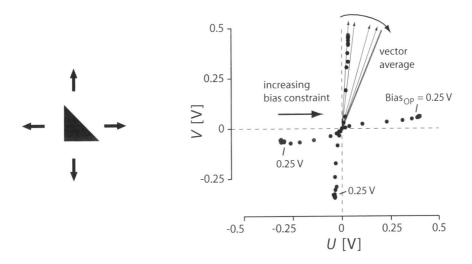

Figure 6.8 *Intersection-of-constraints solution.*
The global optical flow estimate is shown for a visual stimulus containing a triangular object moving in any of the four cardinal directions. The collective computation of the intersection-of-constraints (IOC) solution is required in order to achieve a correct estimate of image motion. Data points along each response trajectory correspond to particular bias voltages Bias$_{OP}$ (0, 0.1, 0.15, 0.2, 0.25, 0.3, 0.33, 0.36, 0.39, 0.42, 0.45, 0.48, 0.51, 0.75 V) where increasing voltages decrease the length of the estimated motion vectors. For this stimulus, the smooth optical flow chip is capable of solving the aperture problem.

estimate approximates the direction of the vector average and the reported speed decreases rapidly. At high bias voltages Bias$_{OP}$ the bias constraint becomes dominant and the chip only reports the reference motion (U_{ref}, $V_{ref} = 0$). The chip's behavior matches the theoretical predictions very well. An increase in bias strength is qualitatively equivalent to a decrease in stimulus contrast. Thus the recorded trajectories fit the simulations for decreasing contrast as shown in Chapter 4 (see Figure 4.6). Note that in practical applications a minimal level of Bias$_{OP}$ is always needed in order to counter-balance any second-order effects and is determined as the minimal voltage for which the chip still robustly reports a zero-motion output if the visual contrast vanishes and the photoreceptors are completely adapted.

6.3 Flow Field Estimation

The basic characterization of the smooth optical flow chip was performed on a set of well-defined artificial stimuli. Also, the previous measurements represented the global optical flow vector, averaged over one stimulus cycle. This, of course, does not reflect the typical perceptual mode of, for example, an autonomous robot behaving in a dynamic real-world environment. The optical flow chip, however, is limited neither to artificial stimuli, nor to a global read-out mode. It contains scanning circuitry that permits the (almost) instantaneous read-out of the photoreceptor and the optical flow signals of each pixel (up to 5000 frames/s). In the following, I will give a few such instantaneous optical flow estimates for some natural and artificial visual stimuli. This time, the data are obtained from the optical flow sensor containing the larger array (30×30 pixels).

Figure 6.9 shows the scanned output (photoreceptor signal and optical flow estimate) for various stimuli, such as moving dark dots on a light background (a), a black and white grid moving horizontally to the left (b), and part of a waving hand on a light background (c). For the above examples, the lateral coupling between units (smoothness constraint) was fairly limited. Consequently, the resulting flow fields are smooth, yet provide a sufficiently local estimate of optical flow. The estimated optical flow tends to exhibit normal flow characteristics in areas where integration on a larger spatial scale would be needed in order to resolve the aperture problem. This can be observed for the grid pattern where the optical flow vectors have the tendency to be perpendicularly oriented to the edges of the grid (Figure 6.9c). Only at the grid nodes are spatial gradients of different orientations provided within the local kernel of interaction which makes an IOC solution possible.

Figure 6.10 demonstrates how different bias values of the HRes circuits can control the smoothness of the optical flow estimate. The same "triangle" stimulus as in Figure 6.8 was presented, moving from right to left with constant speed. Optical flow was recorded for three different strengths of smoothness (Bias$_{HR}$ = [0.25, 0.38, and 0.8 V]) and the chip's output is shown at times when the object approximately passed the center of the visual field. Figure 6.10a is a normalized gray-scale image of the photoreceptor output voltages while b–d display the estimated flow fields for increasing smoothness strengths. For low values, spatial interaction amongst motion units was limited and the result is a noisy normal flow estimate with flow vectors preferably oriented perpendicular to the brightness edges. As the voltage increases, the optical flow estimate becomes smoother and finally represents an almost global estimate that approximates well the correct two-dimensional object motion.

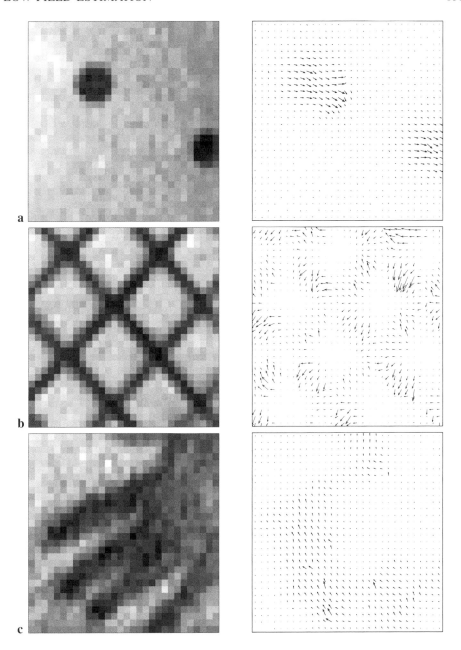

Figure 6.9 *Single frames of the scanned output of the smooth optical flow chip for various "natural" stimuli.*
The scanned photoreceptor signals and the estimated optical flow field are displayed for (a) moving dark points on a light background, (b) a grid pattern that moves to the left, and (c) a hand moving in the direction of the upper left corner.

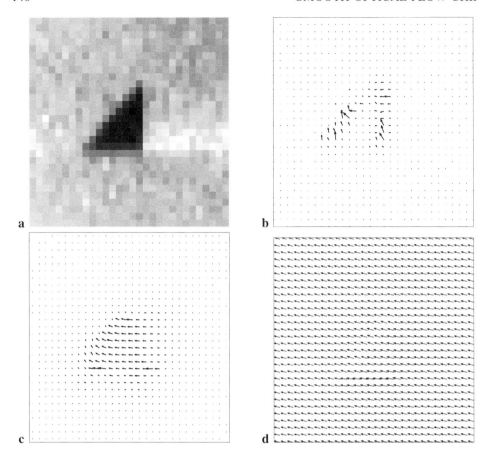

Figure 6.10 *Optical flow estimates of increasing smoothness.*
(a) The photoreceptor signal represented as a gray-scale image. (b–d) Sampled optical flow
estimates for increasing weights of the smoothness constraint (Bias_{HR} = [0.25, 0.38, and 0.8
V]). The output continuously changes from a normal flow toward a global IOC estimate.
(Analog Integrated Circuits and Signal Processing, Volume 46, No. 2, 2006, Analog integrated 2-D optical flow sensor, Stocker, A. Alan, with kind permission of Springer Science
and Business Media.)

The last example illustrates how the smooth optical flow sensor behaves for "natural"
real-world scenes. In this case, "natural" refers to a typical indoor office environment.
Figure 6.11 shows a sampled sequence of the optical flow sensor's output while it is
observing two tape-rolls passing each other on an office table from opposite directions.[2]
The shown frame-pairs (a–f) are in chronological order with an interframe time difference

---

[2]This scene was intentionally selected in order to refer to the "tape-rolls" sequence used in the analysis of the
optical flow network in Chapter 4. It is left to the reader to compare the simulation results to the responses of the
smooth optical flow chip.

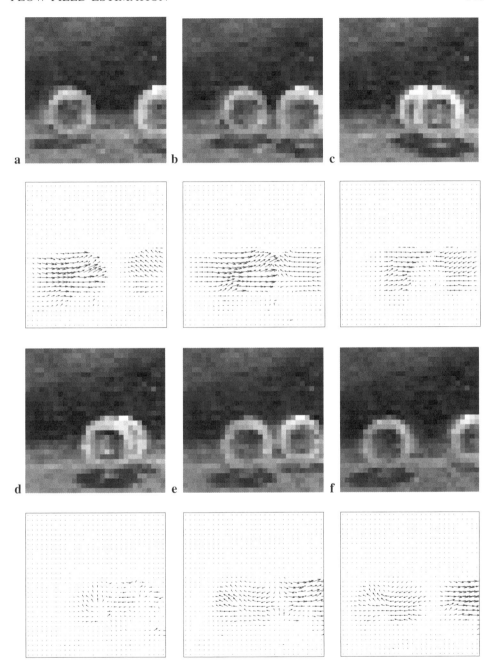

Figure 6.11   *Sampled output of the smooth optical flow chip perceiving a real-world scene.* This and other additional sequences are available as full movies on http://wiley.com/go/analog.

of $\Delta t \approx 50$ ms. Each sample pair consists of two images: a gray-scale image that represents the photoreceptor outputs and, below, the estimated optical flow field. Despite the relative low resolution, the action in the sequence can be well reconstructed. The two tape-rolls are crossing each other in the fourth frame, where the roll that started from the left is in the back. This roll is also moving faster than the other roll as can be recognized from the interframe displacements – and which is correctly reflected in the length of the estimated flow vectors (see frames a, b, and f). Overall, the optical flow estimate matches well the qualitative expectations. Rotation of the rolls cannot be perceived because the spatial brightness structure on their surface is not sufficient. The flow fields are mildly smooth ($Bias_{HR} = 0.41$ V) leading to motion being perceived also in the center-hole of each roll, for example. The chosen value for the smoothness conductance permits the sensor to estimate the perceptually correct object motion in the middle of the tape-rolls, whereas at the outer borders of the rolls, the smoothness kernel is not large enough to include areas with spatial brightness gradients of sufficiently different orientations. The aperture problem holds and the estimated optical flow field tends toward a more normal flow estimate.

The tape-rolls sequence demonstrates another interesting behavior of the sensor: the optical flow field seems partially sustained and trails the trajectory of the rolls (see e.g. the optical flow field on the left-hand side in Figure 6.11c), indicating a relative long decay time constant of the optical flow signal. This can be explained by considering the total node conductances in the network as a function of the visual input. The total conductance at each node of the optical flow network is mainly given by the transconductance of the feedback loop and the bias conductance (see Equation (5.30) on page 123); for reasons of simplicity, we neglect the influence of the smoothness and the intrinsic (parasitic) bias conductance. The transconductance of the feedback loop, thus the feedback currents as a function of the voltages U_+ and V_+, is directly weighted with the squared magnitude of the spatial brightness gradients (Equation (4.23) on page 57). The transconductance is high at high-contrast edges of the tape-rolls. Once the rolls have passed, the unstructured background reveals almost no spatial brightness gradients, and the transconductance becomes low. Consequently, the time constant increases because it is now determined solely by the bias conductance ρ – which is ideally small. As a counter-example, consider the situation after the occlusion of the two tape-rolls: there is almost no trailing expansion of the flow field right after occlusion (Figure 6.11e) because the high-contrast back edge of the second roll immediately induces new motion information. A static background could provide motion information (zero motion) as long as it contains sufficient spatial brightness gradients. This is, however, not the case in this scene. In the last frame (Figure 6.11f) the trailing flow field is already growing again. The perceptual consequences of these dynamical effects will be discussed in more detail in Section 6.5. As expected from its implied smooth motion model, the optical flow sensor cannot resolve occlusions where multiple motion sources are present in the same spatial location. In this particular sequence with opposite translational motion sources, the motion estimates cancel at occlusions (see Figures 6.11c and d).

6.4   Device Mismatch

Analog VLSI circuits are affected by noise from different sources, one of which is fabrication "noise" or mismatch. Minimal misplacements of different lithographic masks can cause geometrical deviations from the designed layouts. Also inhomogeneities in doping and simply unavoidable random variations throughout the complete fabrication process are the reasons why two circuit elements, although laid out identically by the designer, will never behave exactly identically on-chip. These variations might be minimized due to rigorous process control but they will always be present and will always constrain analog circuit design. It is important to investigate how these circuit variations are manifested in the computational behavior of the smooth optical flow chip. And, of course, it is of interest to know how strong theses effects are for the particular design.

6.4.1   Gradient offsets

It turns out that local offsets in the computation of the spatial and temporal brightness gradients can account to a large degree for the deviations from the ideal optical flow estimate. Additive and multiplicative offsets thereby affect the estimate differently. To illustrate this, the response of a single optical flow unit to a sinewave brightness pattern was simulated for both types of offset and different offset amplitudes, and compared to the measured output of a single unit of the smooth optical flow sensor.

As analyzed in Section 5.3, the motion response of a single unit to a moving sinewave brightness grating is expected to be phase-dependent due to the bias constraint. Measuring the response of a single, isolated optical flow unit ($\rho = 0$) confirms the expectation. Figure 6.12a shows the measured signal trajectories for a drifting sinewave grating oriented orthogonal to the u direction. The top trace represents the photoreceptor output of the unit. The mean response of the motion output over a complete stimulus cycle is in qualitative agreement with the presented visual motion, having a positive U component and an approximately zero V component. The individual trajectories, however, typically vary substantially during one stimulus cycle. The trajectories are measured from a unit at the array border and are therefore expected to be affected by mismatch slightly more than average.

Figure 6.12b shows the simulated instantaneous response of a single optical flow unit in the presence of offsets in the spatial brightness gradients. The gradients were computed according to (5.8) and (5.9), respectively. Each trace represents the response according to (5.10) with a constant (positive) offset Δ added to the spatial gradient. For $\Delta = 0$, the output reports the correct image motion most of the time, except when the spatial gradient becomes zero and the bias constraint dominates and enforces the reference motion ($V_{ref} = 0$). Adding offsets in the order of 25% or 50% of the maximal spatial gradient amplitude changes the response behavior drastically. When the offset increases the absolute value of the spatial gradient estimate, the motion output decreases. It even changes sign in the interval $E_x = [-\Delta, 0]$. Conversely, when the offset decreases the absolute value of the spatial gradient estimate, the motion response is increased. The amount of increase and decrease in response is not equally large, due to the divisive influence of the spatial gradient in the computation of the motion output (5.10). The output becomes increasingly asymmetric with increasing offsets. The locations of the extrema in the simulated trajectories undergo a phase shift which can be observed in the measured trajectories as well.

The effective additive mismatch in the spatial gradient computation on the smooth optical flow chip was estimated by measuring the variations of the individual photoreceptor outputs in their adapted state to a uniform visual input. These variations lead to an average additive offset in spatial gradient estimation of 21 ± 18 mV (standard deviation).

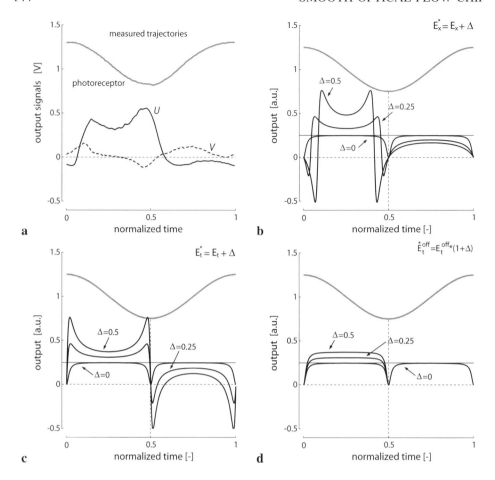

Figure 6.12 *Offset in the extraction of the brightness gradients.*
(a) The output trajectories of a single isolated optical flow unit to an orthogonal oriented drifting sinewave grating exhibit non-ideal behavior. The photoreceptor output (top trace) and the optical flow components U and V are shown. Ideally, $V = 0$ and U is symmetric in both half-cycles of the stimulus. Offsets in the computation of the spatial and temporal brightness gradients due to mismatch in the photoreceptor and in the hysteretic differentiator circuits can explain the observed behavior. (b) Additive offsets in the spatial gradient computation result in an asymmetric and phase-shifted motion output. (c) Additive offsets in the temporal gradient estimation also induce asymmetric changes in the response but no phase shift. (d) A gain offset in the off-transient (or on-transient) current of the hysteretic differentiator results in an asymmetric response that significantly affects the time-averaged response. Gain offsets are likely to be induced by mismatches in the current mirrors of the differentiator circuit.

The relative impact of this offset depends of course on the maximal spatial gradient signal imposed by the stimulus. For example, the applied sinewave grating stimulus in Figure 6.12a had a peak contrast of 80% which leads to a peak-to-peak amplitude of the photoreceptor output of approximately 500 mV. Given a spatial stimulus frequency of 0.08 cycles/pixel, the maximal local spatial gradient signal is approximately 125 mV. Measured additive offset is in the order of 17% of the maximal effective spatial gradient. The photoreceptors, however, are obviously not adapted when exposed to a time-varying stimulus. Thus, the additive mismatch will be superimposed with mismatch in the transient gain of the photoreceptor circuits. The transient gain is given by the capacitor divider ratio of the photoreceptor circuit. Although the matching of poly–poly capacitors is typically much better than matching transistor sizes, the gain mismatch will certainly increase the overall photoreceptor offset beyond the measured 17%.

The additive offsets in the temporal gradient computation have a similar effect on the resulting output, which is illustrated in Figure 6.12c. However, the responses to both stimulus half-cycles are more symmetric than with offsets in the spatial gradient. The phase shift of the peak responses does not occur. Additive offsets can occur due to leakage currents in the source implants between the native and the well transistor in the rectifying element of the hysteretic differentiator circuit (see circuit diagram in Figure 5.3 on page 101).

A third possible source of mismatch consists of the gain differences between the rectified output current of the hysteretic differentiator for positive (on-transient) and negative (off-transient) brightness changes. Figure 6.12d depicts the expected motion output for an increased gain in the off-transient. Although the effect is much smaller compared to the additive effects discussed before, gain offsets amplify the effects of the additive offsets in computing the gradient. It is very likely that such deviations of gain occur in the temporal gradient computation: the on- and off-currents of the hysteretic differentiator pass through several current mirrors before entering the feedback loop, and the transistor mismatch in each current mirror circuit contributes to the total gain offset.

6.4.2 Variations across the array

It is interesting to investigate the output variations across all units of the smooth optical flow chip. To do so, a diagonally oriented moving sinewave grating was applied such that both components (u,v) of the normal flow vector were equally large in amplitude. Each unit of the sensor was recorded in isolation with the smoothness constraint being completely disabled (Bias$_{HR}$ = 0 V). The stimulus speed and the bias settings were chosen such that the outputs were clearly within the linear range of the optical flow units to prevent any distortion of the output distribution due to the expansive non-linearity of the circuit.

Figure 6.13a presents the resulting time-averaged (20 stimulus cycles) optical flow field estimate. It clearly shows the variation in the optical flow estimates of the different units in the array. Units that exhibit the largest deviations seem to be located preferably at the array boundaries where device mismatch is usually pronounced due to asymmetries in the layouts. Because optical flow is computed component-wise, mismatch is causing errors in speed as well as orientation of the optical flow estimate. Figure 6.13b shows the histogram of the output voltages of all motion units. The outputs consistently approximate normal distributions as long as the motion signals are within the linear range of the circuit, which is in agreement with randomly induced mismatch due to the fabrication process. The above

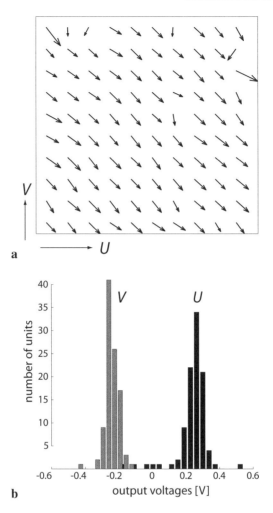

Figure 6.13 *Variations across the array.*
(a) The measured time-averaged response of all isolated optical flow units in the 10×10 array is plotted. The stimulus was a sinewave grating, diagonally oriented and drifting to the lower right corner. (b) The histograms for both components of the optical flow vector quantitatively show the distribution of the individual motion estimates. Mean values (plus/minus standard deviations) are $\overline{U} = 0.25 \pm 0.09$ V and $\overline{V} = -0.24 \pm 0.04$ V respectively. (Analog Integrated Circuits and Signal Processing, Volume 46, No. 2, 2006, Analog integrated 2-D optical flow sensor, Stocker, A. Alan, with kind permission of Springer Science and Business Media.)

offset values were measured for completely isolated motion units. Very similar results were found for other chips of the same fabrication run. Already weak coupling among the units ($\text{Bias}_{HR} > 0$) noticeably increases the homogeneity of the individual responses. It is interesting to note that offsets in the computation of the temporal brightness gradient do not induce directional variations of the optical flow field. Because both flow components are proportional to E_t (see (5.10) for the case $\sigma = 0$), these offsets only change the length of the optical flow vector.

6.5   Processing Speed

Although the output of the smooth optical flow chip is analog and time-continuous, this does not imply that the output always represents the optimal optical flow estimate. As discussed earlier,[3] the instantaneous output of the smooth optical flow chip can be assumed to be the optimal solution of the optimization problem, if the time constant of the implemented network is negligible compared to the dynamics of the visual input. The time constant provides a processing speed limit beyond which the optical flow estimate eventually deviates largely from the optimal solution of the estimation problem. If the dynamics of the visual scene exceed the time constant of the network, the global minimum can change its location faster than the network is able to reduce the distance to it and thus it may never reach it.

The time constant of the smooth optical flow chip is mainly determined by the capacitances C_u, C_v and the total conductance at the network nodes of each individual optical flow unit. C_u and C_v are not explicitly designed but represent the sum of the various parasitic capacitances present. They are typically very small and might vary slightly between different optical flow units. The total conductance at each network node consists of the various conductances attached, such as the transconductance of the feedback loop, the transconductance of the amplifier circuit that implements the bias constraint, the lateral smoothness conductances, and also the output conductance at the feedback node. As discussed before, all these conductances are non-ohmic.

A detailed and exact analysis of the dynamics is complicated by the non-linearities of the different conductances. Even when the smoothness conductance is neglected and the remaining conductances are assumed to be linear, it is still necessary to solve the inhomogeneous system of coupled differential equations (see Equation (4.21)) and a clear time constant would be hard to define. Instead, for the purpose of providing a more intuitive understanding of the dynamics, a simplified optical flow network under particular stimulus conditions is considered. The analysis is restricted to a single optical flow unit (neglecting the smoothness conductances) and the output conductance of the feedback loop is assumed to be negligibly small. Then, according to the dynamics of each network node (4.21), the total node conductance reduces to the sum of the bias conductance g_{bias} and the transconductance of the feedback loop g_{fb}. Furthermore, the stimulus is assumed to be essentially one-dimensional (or a superposition of one-dimensional stimuli) such that the brightness gradient in one direction vanishes. Under this assumptions, the dynamics become separable and the time constant can be easily determined. In the following, this is done exemplarily for one optical flow component. For the u component of the optical flow, for example, the transconductance of the feedback loop g_{fb} is given as the amount of change in feedback current for a change in the node potential U (see Figure 5.8), thus

$$g_{\text{fb}} = \frac{\partial}{\partial U}(Fu_- - Fu_+). \qquad (6.1)$$

---

[3]See Section 4.2.1 on page 57.

The feedback currents are ideally proportional to the partial gradients of the brightness constancy constraint (4.23) and therefore the transconductance

$$g_{fb} \propto \frac{\partial F_u}{\partial u}$$
$$= \alpha(E_x)^2, \tag{6.2}$$

where α is a constant that accounts for the variable current gain of the wide-linear-range multiplier circuit defined by the bias voltage $Bias_{VVI2}$.

With the above assumptions, the time constant τ is given as

$$\tau = \frac{C_u}{g_{fb} + g_{bias}} = \frac{C_u}{\alpha(E_x)^2 + g_{bias}}. \tag{6.3}$$

The time constant is proportional to the node capacitance C_u that is inherently given for a particular implementation. However, the current gain of the wide-linear-range multiplier α allows us to adjust the time constant to some degree and so does the bias conductance. Yet, even for a given set of these parameters, the time constant and thus the processing speed are not constant but *stimulus-dependent*. They vary inversely with the squared amplitude of the spatial brightness gradients. To verify this, some measurements of processing speed were performed. As illustrated in Figure 6.14, the smooth optical flow chip was presented with a plaid stimulus composed of two orthogonal sinewave gratings. The spatial and temporal frequencies of each grating were identical. The plaid pattern was presented such that the orientation of each grating was parallel to one of the motion component axes. While the contrast c_1 of the first grating was varied $[0 \dots 50\%]$, the contrast of the second one was kept constant ($c_2 = 50\%$). Processing time was defined as the time needed by the chip to raise the voltages U and V to 75% of their asymptotic signal levels after onset of the stimulus. Before this onset, the visual input to the chip was a uniform surface of constant brightness. Although the chip was biased to provide the global optical flow estimate, the global visual stimulus presumably makes the measured time constants comparable to the time constant of an individual optical flow unit. The measurements in Figure 6.14 show how processing time depends on contrast. If both gratings are of high contrast, the processing time is approximately 30 ms and equal for both components of the optical flow estimate. As the contrast of the grating orthogonal to the u direction decreases, the processing time for the u component strongly increases. At zero contrast, no visual information is present in that direction and the measured processing time is determined only by the bias conductance.[4] Because of the logarithmic photoreceptor, contrast is directly equivalent to the brightness gradients. Therefore, the measured processing time as a function of contrast was fitted according to (6.3). The fit shows that the simplified analysis is a valuable approximation.

The above example represents the step response of the smooth optical flow chip. In this case it is sensible to consider the time constant as an inverse measure of processing speed. Under more natural conditions, however, it makes more sense to consider it as the characteristic time constant of motion integration, because the visual input is continuously updated while the chip computes the estimate. This is in contrast to the more traditional

---

[4]The error bar is evidently huge because the component's value is ideally constant and zero and the 75% limit only determined by noise.

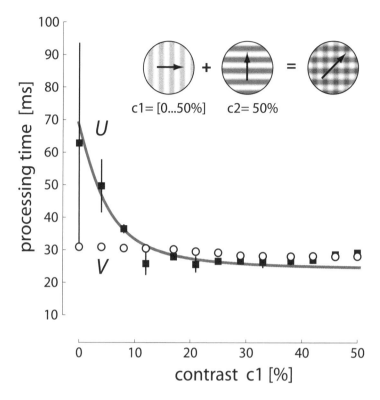

Figure 6.14 *Processing speed of the smooth optical flow chip.*
The graph shows the time needed for the output voltages U and V to reach 75% of their asymptotic output level after the onset of a moving plaid stimulus (30 pixels/s). The plaid consisted of the superposition of two sinewave gratings (spatial frequency 0.08 cycles/pixel) where the contrast of the grating oriented perpendicular to the u component was varied while the contrast of the other grating remained constant. Each grating only drives the dynamics for the optical flow component orthogonal to its orientation. Processing time is proportional to the time constant and the fit is according to (6.3). Data are recorded from the small array.

meaning of processing speed, which relates to the time needed to find a solution for a *given* input (one frame). The perceptual impact of the contrast-dependent temporal integration under natural visual stimulation has already been noted for the flow field estimations shown in Figure 6.11.

- *Low spatial contrast:* If the local stimulus contrast is very low, the spatio-temporal brightness gradients vanish and the feedback currents at the capacitive nodes are small. The time constant becomes large because the conductance is given mainly by the transconductance of the amplifier and the remaining output conductance of the cascode current mirror, which are both ideally small. The lower the visual contrast, the lower the processing speed. This is a sensible property, because in this way the smooth optical flow chip can distinguish between high temporal frequency noise of low amplitude and reliable but low spatio-temporal frequency input such as a slowly moving object with shallow spatial brightness gradients. In both cases the time constant is long. In the first case, the long temporal integration averages out the noise and the chip does not elicit a noticeable output signal. In the latter case, however, the long time constant provides the necessary integration time to build up a reliable motion estimate.

- *High spatial contrast:* Conversely, high local stimulus contrast increases the feedback currents and leads to faster processing in the smooth optical flow chip. This is also a sensible property because locations of high spatial contrast typically represent reliable stimulus features (*e.g.* edges) that need a shorter integration time and thus allow a higher temporal resolution.

The optical flow network intrinsically adapts its processing speed to the confidence it has on the visual input it receives. The combination of the bias constraint and the adaptive time constant of temporal integration makes an explicit confidence measure obsolete. Furthermore, the adaptation of the time constant is a local process. Each individual optical flow unit adapts its integration time constant individually according to the confidence in its own local visual input. Such a locally adaptive scheme is better than a global one because it adapts perfectly to any visual input.

6.6   Applications

In this last section, I will illustrate how the smooth optical chip can be used in applications, despite the fact that the current implementations are only prototypes with very limited array resolution. Clearly, the smooth optical flow chip is not designed for precise velocity measurements as required, for instance, in industrial process control. Applications under real-world visual conditions, however, are the domain in which the chip can reveal its true potential; the motion model with its bias constraint provides the necessary robustness of the estimate and the adaptive phototransduction stage permits operation over many orders of light illumination.

In some way, applications also represent the ultimate test of the motion model and the aVLSI design of the smooth optical flow chip. In a real-world environment, the visual conditions are typically not as controlled as in the laboratory and the noise level can be significantly higher. For example, light flicker (50/60 Hz) is a dominant noise source that is omnipresent in an indoor environment. Flicker noise induces high temporal brightness gradients that can cause spurious motion signals due to the brightness constancy constraint. Appropriate bias settings of the photoreceptor circuit, however, can change the photoreceptor's band-pass characteristics such that these high-frequency brightness signals are usually sufficiently damped. Other noise sources are directly related to the chip's analog information processing and signal representation. Small voltage fluctuations in the power supplies, for example, directly affect the signals on the chip. Care has to be taken to stabilize the power supply. Also electromagnetic interference originating, for example, from electric motors on a common robotic platform can cause similar problems. Compared to digital circuits, the generally increased noise susceptibility is a downside of analog computation and representation. Nevertheless, the following applications prove that it is not prohibitive. Yet, it requires a careful design of such applications and promotes hybrid analog/digital processing frame-works: local parallel computation can be very efficiently performed in analog circuits as demonstrated by the smooth optical flow chip. Long-range communications and signal storage, however, are more robust when performed digitally.

6.6.1 Sensor modules for robotic applications

Autonomous mobile systems are the natural application platform for the smooth optical flow chip. The chip's efficiency makes it particularly interesting for smaller robots where size and power restrictions severely limit the available computational resources for visual processing. Efficiency becomes extremely important in small flying robots. Visual processing is almost prohibitively expensive in terms of power, size, and weight restrictions on such platforms. Visual motion information is useful for a flying robot for several reasons. A sudden change in the optical flow can indicate an obstacle, and should trigger an appropriate flight maneuver to avoid collisions. Motion information from optical flow can also be used to stabilize the flight trajectory. Honeybees, for example, align their flight trajectories when flying through narrow channels or tunnels such that the optical flow fields on both lateral hemispheres match [Srinivasan et al. 1991]. Because the magnitude of optical flow is to a first approximation inversely proportional to the lateral distance between the bee and an object or wall, matching the flow fields ensures that the bee is roughly flying in the middle of the channel or tunnel, thus is on average furthest away from any obstacle. Honeybees also seem to measure and integrate optical flow along a flight trajectory in order to obtain a measure for flight distances; when honeybees discover an attractive food source they communicate to their nestmates the location and distance of the source relative to the hive by the famous "wiggle-dance". However, they clearly under-estimate the distance when the flight leads through terrain that induces only weak optical flow, such as water surfaces [Tautz et al. 2004].

Applications like these in artificial flying agents require sensor modules of minimal weight. A first prototype module based on the smooth optical flow chip has been developed and is shown in the pictures of Figure 6.15. The chip is set to compute the global 2-D optical flow vector. As seen in Figure 6.15a, the silicon die is directly bonded onto the printed circuit board (PCB) saving the weight of the chip package. The weight of necessary peripheral electronics (e.g. the potentiometers for setting the bias voltages on-chip) is kept minimal using surface-mounted devices (SMD).

A global optical flow estimate can be sufficient for many applications. It has the advantage that the output signal is of very low dimensionality and that the scanning circuitry of the smooth optical flow chip does not need to be activated, which would require some external clock signal. Reading out the signals of the entire array of the smooth optical flow chip, however, offers even more sophisticated processing possibilities. Not only the optical flow field but also the brightness distribution can be obtained from the chip. Thus a complete low-resolution *vision module* based on the smooth optical flow chip can take full advantage of its efficient analog processing. Such a vision module is currently under development [Becanovic et al. 2004]. It combines the smooth optical flow chip with a microprocessor that primarily handles the read-out of the chip, performs the digital conversion of the data, and provides standard interfaces to allow an easy and seamless integration of the module within complete sensory-motor systems. In addition, the microprocessor can

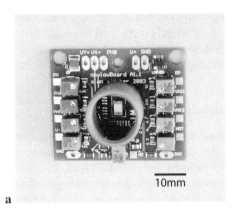
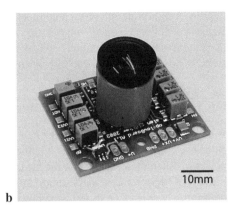

a   b

Figure 6.15 *Compact visual motion module for robotic applications.*
(a) The silicon die is directly bonded onto the printed circuit board (PCB) in order to save
the weight of the chip package. Only a few peripheral components are needed, such as
some micro-potentiometers for setting critical bias voltages. The module's outputs are two
analog voltages representing the global 2-D optical flow estimate. (b) The module with
mounted wide-angle lens. This first prototype weighs only 6.2 g without the lens (11.4
g with a glass lens) and measures 36 mm by 30 mm by 20 mm in size. Using plastic
lenses and optimizing components and wiring densities will permit a further reduction of
the module's weight and size.

be used to postprocess the optical flow and image brightness data from the chip. Different
tasks can be programmed such as the extraction of heading direction (focus-of-expansion)
or motion segmentation. Such a vision module could provide relative cheap yet powerful
vision capabilities to small robotic platforms.

6.6.2   Human–machine interface

In human interactions there occur many situations where information is conveniently encoded and transmitted via short sequences of visual motion. For example, we nod our
heads to indicate agreement, and shake them to signal disagreement; We greet somebody
at a distance by waving our hand, or to say "goodbye". Such forms of communication are
in general useful to gain attention and when oral communication is not possible.

We developed an artificial system that can recognize simple sequences of visual motion
for human–machine interaction [Heinzle and Stocker 2003]. The computational architecture of the system is entirely based on analog constraint satisfaction networks although its
implementation is hybrid (analog/digital). The visual motion frontend consists of the smooth
optical flow chip estimating global 2-D visual motion. The classification, however, is performed by two analog networks that are emulated on a digital computer. A hybrid system
combines the strengths of both domains. The chip computes global visual motion efficiently
and reduces the computational load for the digital computer to such a degree that it can
comfortably emulate the classification networks in real time in a high-level programming
language. The digital backend, however, provides high programming flexibility.

Figure 6.16a shows the complete system. The smooth optical flow chip is connected to the digital computer via a universal serial bus (USB) interface. Using a USB interface has the advantage that the intrinsic power line of the bus can be used as the supply voltage for the chip, which is convenient and facilitates mobile applications. The smooth optical flow chip with USB interface is shown in more detail in Figure 6.16b. It basically consists of two stacked PCBs. The first contains the smooth optical flow chip while the second hosts a small microcomputer with digital/analog converter that reads and converts the global optical flow signal, and additionally contains some generic driver circuits that handle the serial communication over the USB interface.

Classification of the motion sequences is achieved in two processing steps performed by competitive analog networks emulated on the digital computer. These two networks

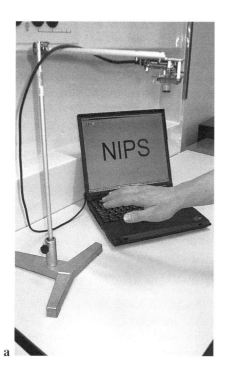

a   b

Figure 6.16 *Visual human–machine interface.*
(a) The picture shows the human–machine interface in action. The system consists of the smooth optical flow chip and two constraint-solving classification networks emulated on a laptop computer. It is capable of classifying short sequences of global motion and assigning, in this example, single alphabetic letters to individual sequences of motion. Here, arm movements are detected, yet more subtle motions elicited by, for example, small head movements could be detected and applied. (b) The smooth optical flow chip with optics and universal serial bus (USB) interface. A USB interface provides a convenient way to connect the smooth optical flow chip to a digital computer. Note that using the intrinsic power supply of the USB port makes a separate supply for the chip obsolete and allows for a mobile and easy to use "plug-and-play" device.

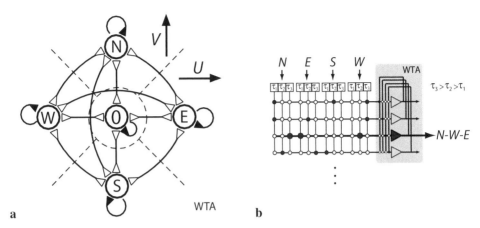

Figure 6.17 *Classifying sequences of global visual motion using analog constraint satisfaction networks.* (a) The *direction selective network* transforms the continuous 2-D global motion estimate (provided by the two voltage signals U and V) into motion events. Motion is classified into the four cardinal directions (plus zero motion). The network architecture contains five units and performs a dynamical soft-WTA competition. (b) The *sequence classification network* classifies sequences of different time-delayed combinations of motion events also using a soft-WTA network architecture. The number of time delays is defined by the length of the sequences to be classified, and the number of network units is equivalent to the number of different sequences to classify.

are basically equivalent to the WTA networks discussed in Chapter 3. First, motion events are detected and extracted by the *direction selective network* depicted in Figure 6.17a. A motion event is defined as global motion in one of the four cardinal directions. The global motion space is divided into five subregions. Each subregion is represented by a single unit of the network: four for the cardinal directions (N–E–S–W) and one for zero motion. Each unit gets input whenever the global motion estimate lies in its subregion, which then eventually makes this unit win the competition. Although only one unit gets input at a time, a competitive network is sensible. The time constant of the network acts as a low-pass filter and thus integrates global motion over some temporal window. More important, hysteresis can be forced by setting the connection weights appropriately (see self-excitation gain limit in Chapter 3). Hysteresis increases robustness insofar that a unit is winning the competition only if it received global motion that is "substantially" different from the motion represented by the previously winning unit. This avoids a randomly oscillating winner selection that otherwise can occur when the motion signal moves within the boundary area in between subregions.

The direction selective network transforms the continuous trajectory of the global optical flow vector into activation profiles of its five units. The outputs of the four directional units (N–E–S–W) are the inputs to the second network, the *sequence classification network* (Figure 6.17b). Each output is fed in parallel into different delay elements, where the number of elements is defined by the length of the motion sequences to be classified (in this case three). A unique combination of delayed outputs form the input to one unit of the classification network. Thus if a certain sequence of motion events matches the particular combination of delayed outputs, the information is concentrated in time and produces a large input. There are as many units as there are numbers of distinct sequences to recognize. The sequence classification network is a competitive network (soft-WTA) such that only one unit can win, thus only one sequence is recognized at a time. In Figure 6.17b, for example, the particular sequence N–W–E of motion events matched the combination of delayed filters of the third classification unit and made it win the competition.

This system has been successfully used as a *virtual typewriter*. Distinct sequences of three motion events were assigned to single letters of the alphabet plus some extract functions ("delete", "shift", etc.). Figure 6.16a shows the setup: sequences of hand movements were sensed by the smooth optical flow chip, classified by the analog networks emulated on the laptop, assigned to letters, and finally displayed directly on the computer screen. The virtual typewriter has proven to be surprisingly robust with respect to varying illumination conditions as well as different human users. The system tolerates time-warps up to 50% before it starts to miss a motion sequence. The smooth optical flow chip contributes substantially to the robustness of the whole system, as well as the relative coarse compartmentalization of the motion space into only five subregions.

7

Extended Network Implementations

In this chapter, I introduce two sensors that are extensions of the smooth optical flow chip. Following the recurrent network architectures described in Chapter 4, the *motion segmentation chip* and the *motion selection chip* provide the dynamic adaptation of the local smoothness and bias conductances, respectively. The pixel schematics of both sensors are based on the schematics of the optical flow unit. Yet, each sensor follows a different strategy to implement the recurrent network extensions. The motion segmentation chip incorporates the second network within the same focal-plane array on-chip, and so has a very complex pixel design. The motion selection chip, however, only adds the requisite interface and programmability to the optical flow network so that individual bias conductances can be dynamically set in one of two states. The idea is that the recurrent feedback loop is formed with a second, external processor. This digital/analog hybrid approach provides more flexibility in emulating different recurrent network architectures. Both sensors are described in detail and their performance is illustrated with several real-world visual experiments.

7.1 Motion Segmentation Chip

The motion segmentation chip is an attempt to implement the motion segmentation network architecture proposed in Chapter 4 (see Figure 4.13). As shown there, this architecture recurrently connects the optical flow network with a second, two-layer *motion discontinuity network* with units P_{ij} and Q_{ij}. Active units represent active line segments oriented in the x and y direction, respectively, which restrict the spatial integration of visual motion. The motion segmentation chip follows the same recurrent architecture although its motion discontinuity network is simpler; there is no oriented soft-WTA competition imposed on the discontinuity units in order to favor a spatially continuous formation of the active line process. Although this simplification decreases the performance, it was necessary in order to reduce the network connectivity. The high wire density would have exceeded the feasible complexity level of a focal-plane chip architecture. Even in this simplified version, the motion segmentation chip possesses one of the largest and most complex pixels known.

Without the soft-WTA constraints, the cost function for the motion discontinuity network (see Equation (4.38) on page 79) reduces to

$$H_P = \sum_{ij} \left(\alpha P_{ij} + \beta (1 - P_{ij})(\Delta^y \vec{v}_{ij})^2 + \gamma (1 - P_{ij})(\Delta^y \vec{F}_{ij})^2 + 1/R \int_{1/2}^{P_{ij}} g^{-1}(\xi) \, d\xi \right).$$

(7.1)

Accordingly, gradient descent on this cost function leads to the following modified dynamics:

$$\dot{p}_{ij} = -\frac{1}{C_P} \left[\frac{p_{ij}}{R} + \alpha - f(\Delta \vec{v}_{ij}^y) - h(\Delta \vec{F}_{ij}^y) \right].$$

(7.2)

The quadratic norm in (4.39) has been replaced by the more general functions f and h, which will later be specified for the particular aVLSI implementation. The vector $\vec{F} = (Fu, Fv)$ is composed of two components that represent the feedback currents generated by the optical flow unit for each optical flow component as given by (5.14). It is easy to see that each discontinuity unit basically performs a dynamic threshold operation: the input p_{ij} increases as long as the weighted sum of the optical flow gradient and the gradient on the derivative of the brightness constancy constraints is larger than the threshold α plus the leakage term determined by the conductance $1/R$. Analogously, p_{ij} decreases if the input is below α. Assuming a sigmoidal activation function $g : p_{ij} \rightarrow P_{ij}$ that has an upper and lower bound, and $1/R$ to be small, then – in steady state – the output P_{ij} of the discontinuity unit approximates a binary line process.

The recurrent loop is closed by the discontinuity units P_{ij} controlling the smoothness conductances between neighboring units in the y direction (see Equation (4.42) on page 83) of the optical flow network. Equivalently, the dynamics of the units Q_{ij} are defined as

$$\dot{q}_{ij} = -\frac{1}{C_Q} \left[\frac{q_{ij}}{R} + \alpha - f(\Delta \vec{v}_{ij}^x) - h(\Delta \vec{F}_{ij}^x) \right].$$

(7.3)

7.1.1 Schematics of the motion segmentation pixel

Figure 7.1 shows the schematics of a single pixel of the motion segmentation chip. The core consists of the optical flow unit (represented as a black box), and is identical to the circuit diagram in Figure 5.8. The additional circuitry contains the two discontinuity units that control the state of the line process in the x and y directions. The discontinuity units need to receive the appropriate optical flow and feedback current signals from two neighboring pixels, but the pixel must also provide the very same signals to the discontinuity units of the other two neighboring pixels. These connections are the main reason for the very high wire density in the layout. Signals are all represented as voltages. Hence, the feedback current signals $Fu_\pm$, $Fv_\pm$ are logarithmically encoded as the gate voltages $\tilde{F}u_\pm$, $\tilde{F}v_\pm$ of the cascode current mirrors at the feedback nodes (see Figure 5.8).

The function $f(\Delta \vec{v})$ is implemented as the sum of the "anti-bump" outputs of two bump circuits [Delbrück 1991]. The "anti-bump" output current approximates the quadratic norm of $\Delta \vec{v}$. Although each component of the optical flow vector is encoded as a differential signal, the choice of a common virtual ground V_0 only requires comparison of the positive

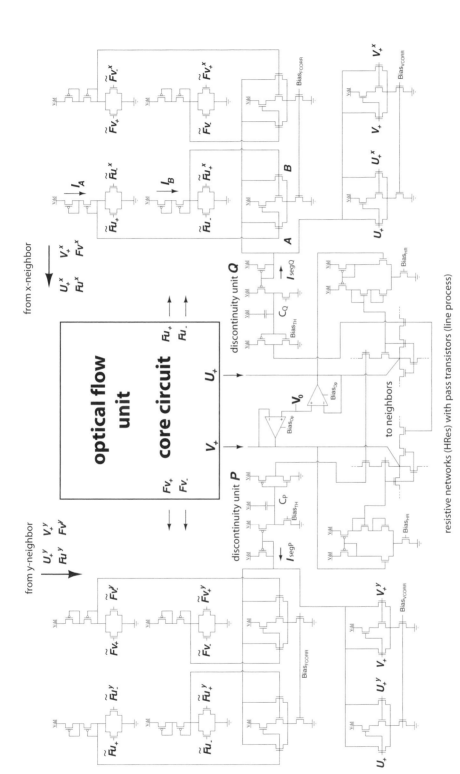

Figure 7.1 *Schematics of a single pixel of the motion segmentation chip.*
The core circuit consists of the optical flow unit as shown in Figure 5.8 (page 110). (Stocker, A. Analog VLSI focal-plane array with dynamic connections for the estimation of piece-wise smooth optical flow, IEEE Transactions on Circuits and Systems – 1: Regular Papers 51(5), 963–973, © 2004 IEEE)

vector components U_+ and V_+ between the pixel and its neighbors. These signals directly serve as the appropriate differential input voltages to the bump circuits. The "anti-bump" output current of the bump circuit can be approximately expressed as a function of the voltage difference ΔV on its input, given as

$$I_{VCORR} = I_b \frac{\frac{4}{S}\cosh^2(\frac{\kappa \Delta V}{2})}{1 + \frac{4}{S}\cosh^2(\frac{\kappa \Delta V}{2})}, \tag{7.4}$$

where S is the transistor size ratio between transistors in the inner and the two outer legs of the circuit, and I_b is the bias current set by the voltage $Bias_{VCORR}$. The ratio S determines the width of the "bump", and so the voltage range over which the bump circuit approximates the quadratic norm. The circuit design is such that this voltage range approximately matches the linear output range of the optical flow unit. For larger ΔV the current I_{VCORR} begins to saturate. The function $h(\Delta \vec{F})$ is also implemented using bump circuits. However, a preprocessing step is necessary in order to transform the feedback current gradient ΔF into an appropriate differential voltage input for the bump circuits. This is described in the next section.

The segmentation current for each of the two orientations (e.g. I_{segQ}) is the sum of the output currents of the two stages of bump circuits. The segmentation current is the input to the discontinuity units, which are implemented with only five transistors each. The circuits approximate the dynamics (7.2) and (7.3). The weighting parameters β and γ for the segmentation current are defined by the bias voltages $Bias_{FCORR}$ and $Bias_{VCORR}$ and are therefore adjustable. The threshold current α is set by $Bias_{TH}$. The total node current then charges the capacitances C_P, C_Q either up or down. The sigmoidal activation function g is approximated by the inverter, which is naturally limited by the voltage rails. The leak conductance $1/R$ is implicitly given as the combined drain conductances at the threshold transistor. As discussed previously in the context of the cascode current mirror, the leak conductance depends on the threshold current level but is typically high. Finally, the feedback loop is closed by connecting the output of the discontinuity units to the pass transistors that either enable or disable the smoothness conductance in between two neighboring optical flow units as set by the horizontal resistor.

Computing the gradient of the feedback current signal

The bump circuit provides a compact and appropriate way to measure the dissimilarity between two voltage signals. Measuring the gradients of the feedback currents with bump circuits requires an additional circuit stage that transforms the difference between two differential currents into an appropriate differential voltage signal.

Figure 7.2 illustrates the transformation circuit used in the motion segmentation chip. Its task is to transform the current difference $\Delta I = I_A - I_B = (I_{A+} - I_{A-}) - (I_{B+} - I_{B-})$ into the voltage difference $\Delta V = V_1 - V_2$. The differential currents are cross-wise summed into currents I_1 and I_2. Stacked pairs of diode-connected transistors generate the voltages V_1 and V_2. The resulting voltage difference[1] is then

$$\Delta V = \frac{1}{\kappa_p}\left[\gamma \ \log\left(\frac{I_{A+} + I_{B-}}{I_{A-} + I_{B+}}\right)\right], \tag{7.5}$$

---

[1]In units of kT/q.

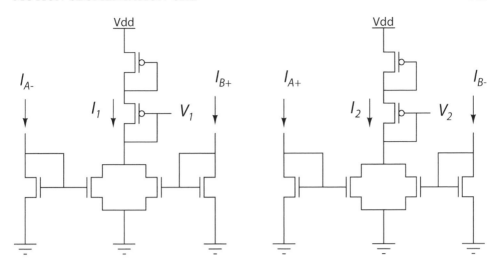

Figure 7.2 *Transforming the difference of two differential currents into a differential voltage.*

where $\gamma = (\kappa_p + 1)/\kappa_p$ with κ_p being the gate coefficient of the diode-connected pFETs. κ_p is assumed to be constant and equal for both pFETs. Assuming that the two differential currents I_A and I_B are symmetric and have a common mean current level I_0, that is $(I_{A+} - I_0) = -(I_{A-} - I_0)$, the difference (7.5) can be linearized around I_0: rewriting the logarithm of the quotient as the difference of two logarithms allows expansion of (7.5) with respect to each differential current component separately,

$$\log(I_{A+} + I_{B-})\Big|_{I_0} \approx \log(2I_0) + \frac{1}{2I_0}[(I_{A+} - I_0) + (I_{B-} - I_0)] - \frac{1}{4I_0^2}[(I_{A+} - I_0)^2$$

$$+ (I_{A+} - I_0)(I_{B-} - I_0) + (I_{B-} - I_0)^2] + R_3$$

$$\log(I_{A-} + I_{B+})\Big|_{I_0} \approx \log(2I_0) + \frac{1}{2I_0}[(I_{A-} - I_0) + (I_{B+} - I_0)] - \frac{1}{4I_0^2}[(I_{A-} - I_0)^2$$

$$+ (I_{A-} - I_0)(I_{B+} - I_0) + (I_{B+} - I_0)^2] + R_3. \tag{7.6}$$

R_3 is the remainder of third-and higher order terms. Taking the difference of the two logarithms in (7.6), the second-order terms cancel because of the symmetry around the common DC level I_0. The remainder R_3 can be neglected when currents I_A and I_B are moderately small compared to I_0, because the components of order n decrease inversely proportionally to I_0 to the power of n. Finally, using (7.6) we can rewrite (7.5) as

$$\Delta V \approx \frac{\gamma}{\kappa_p} \frac{(I_{A+} - I_{A-}) - (I_{B+} - I_{B-})}{2I_0} \propto \Delta I. \tag{7.7}$$

A gain factor $\gamma > 2$ appears because of the additional diode-connected transistor, which enlarges the transformation range. Simulations show that for typical sub-threshold levels of I_0, the useful output voltage range is up to $150\,\text{mV}$, which matches the bump width for the chosen transistor length ratio S.

The linear approximation (7.7) is only accurate under the assumption that the differential currents I_A and I_B have a common DC level. Unfortunately, this is in general not the case in the application here where these currents stand for the feedback currents of two neighboring optical flow units along one direction, for example Fu and $Fu^\times$. The DC level of each feedback current is composed of

- twice half the bias current I_b of the two wide-linear-range multipliers, which is determined by the bias voltage Bias_{VII2} and thus remains constant, and

- a component provided by the hysteretic differentiator which is *not* constant because the total output current of the temporal differentiator circuit is not determined by some bias current but rather depends directly on the visual input.

Typically, the total output current of the hysteretic differentiator is of the same order of magnitude as the multiplier output, thus the variable amount of current it adds to the DC level can be assumed to be in the range $[0, I_b/2]$. Also, the Taylor expansions (7.6) are only good approximations for differential currents that are small compared to I_0. Yet, the output currents of the wide-linear-range multiplier circuits which dominate the feedback currents can be almost as large as their bias currents. Therefore, the transformation circuit in Figure 7.2 only approximates the conversion of the difference of two differential current signals into a proportional voltage difference. In practice, however, deviations are significant only for large-feedback current gradients for which an accurate measure is less critical. The lack of precision in the transformation circuit is at least partially paid off by its simplicity and compactness.

7.1.2 Experiments and results

A fully functional prototype of the motion segmentation chip has been fabricated.[2] The array size of the prototype is small, yet sufficient to provide a proof of concept. It consists of a quadratic array of 12×12 functional units. On-chip scanning circuitry permits read-out of five different output signals from each pixel: the photoreceptor signal, the two optical flow component signals U_+ and V_+, and the output of the two discontinuity units P and Q.

A quantitative evaluation of the motion segmentation chip performance is difficult, because an object-related absolute measure for the quality of the estimated optical flow is hard to define. In computer vision, proposed computational algorithms and architectures are typically tested on representative, but nevertheless somewhat arbitrary, image frames and sequences. Some of these sequences, such as the "Yosemite" sequence, have been used in Chapter 4 for comparisons of the simulated optical flow network to other optical flow algorithms. These sequences essentially serve as quasi-standards. Such standards, however, are not known in the literature on focal-plane visual motion sensors. A fair comparison of aVLSI architectures is often further complicated by the different array sizes of the prototype sensors. My approach will be to illustrate the sensor's behavior by applying two different visual experiments. Due to the small array size, the two experiments include rather simple stimuli. Despite their simplicity, they demonstrate well the advantages of any system that

---

[2]Exact specifications can be found in Appendix D.

preserves motion discontinuities. The experiments are relatively simple to reproduce and can serve to test and compare other approaches.

The test setup was identical to the one applied for the smooth optical flow chip. An optical bench in a dark-room offered controlled conditions for the projection of the stimuli onto the focal plane.

Detecting and preserving motion discontinuities

In the first experiment, the motion segmentation chip was presented with a relatively simple high-contrast visual stimulus. A dark dot was repeatedly moved through the visual field of the chip over a light, uniform background. Figure 7.3 displays a sequence of the scanned output signals of the motion segmentation chip while observing the moving dot. The signals are displayed as two separate images: the top-row gray-scale images represent the photoreceptor signals and are overlaid with the estimated optical flow field. The associated activity pattern of the discontinuity units is shown in the bottom-row images, encoded as gray-scale images where dark pixels represents high local activity in the P or Q units. The activity pattern approximately reflects the outline of the moving dot and thus the location of motion discontinuities. Yet, it seems difficult for the motion segmentation chip to achieve a clean and closed contour around the single object such that the activity of the discontinuity units leads to two fully disconnected areas in the optical flow network. This was expected given the limited model that lacks the soft-WTA competition promoting the formation of contiguous line processes.

A closed contour is desirable. Nevertheless, with every line segment that is active at a motion discontinuity in the visual scene, the quality of the optical flow estimate improves by reducing smoothness across that particular motion discontinuity. Thus, the resulting flow fields should show sharper gradients at the dot's outline. This is indeed the case as shown in Figure 7.4 where estimation results with an enabled or disabled motion discontinuity network are compared. The optical flow estimate was recorded twice for the same moving "dot" stimulus while keeping the default smoothness strength constant (*i.e.* Bias$_{HR}$ constant). The first time, the motion discontinuity network was prevented from becoming active by setting the threshold α to a sufficiently high value. Figure 7.4a shows the corresponding frame of the resulting optical flow estimate and a histogram of the speed distribution in this frame. Ideally, one would expect a bi-modal distribution because the visual scene contains two motion sources, the moving dot and the stationary background. Isotropic smoothing, however, does not preserve motion discontinuities, and the histogram decays monotonically according to a diffusion process. This changes, however, when the motion discontinuity network is activated. Now, the velocity distribution exhibits the expected bi-modal shape as shown in Figure 7.4b.

Visual comparison of the two flow field estimates also reveals the much smoother trail of the optical flow field when the motion discontinuity network is disabled. As discussed previously (see Section 6.3), these smooth trails originate from the long time constant of the optical flow network when the spatio-temporal energy of the visual input is low (as in the unstructured, stationary background). Active line segments separate the trail from the active area and thus speed up its decay over time.

As a side note, the sequence in Figure 7.3 also nicely illustrates the adaptive behavior of the photoreceptor circuit. Because the dot moves repeatedly along the same image trajectory,

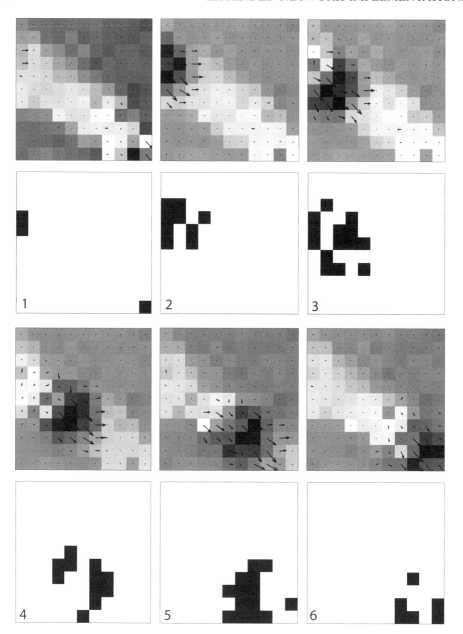

Figure 7.3 *Detecting and preserving motion discontinuities.*
The scanned output sequence of the motion segmentation chip is shown while it was
presented with a moving dark dot on a light background. The signals from each frame are
displayed in two separate images. The photoreceptor outputs with superimposed optical
flow fields are displayed on top of each row. The states of the discontinuity units are
reflected below as a gray-scale image where dark pixels represent active units (P or Q).
(Stocker, A. Analog VLSI focal-plane array with dynamic connections for the estimation of
piece-wise smooth optical flow, IEEE Transactions on Circuits and Systems – 1: Regular
Papers 51(5), 963–973, © 2004 IEEE)

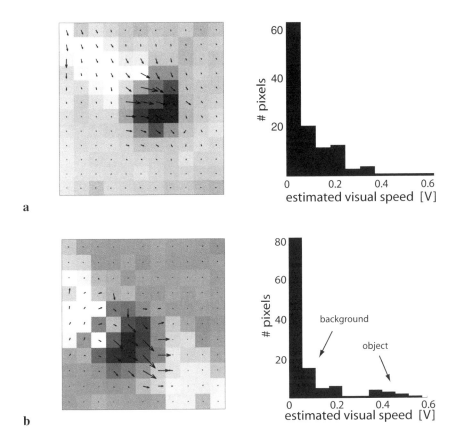

Figure 7.4 *Active line processes reduce smoothing across motion discontinuities.*
Histogram of the optical flow output of the motion segmentation chip with its motion
discontinuity network (a) disabled, and (b) enabled (frame #4 in Figure 7.3). Clearly, the
discontinuity network reduces smoothing across motion discontinuities, which is reflected
in the bi-modal distribution of the speed histogram. (Stocker, A. Analog VLSI focal-plane
array with dynamic connections for the estimation of piece-wise smooth optical flow, IEEE
Transactions on Circuits and Systems − 1: Regular Papers 51(5), 963–973, © 2004 IEEE)

the photoreceptors located outside the dot's path adapt over time to some mean gray value
that represents the logarithmically encoded background light intensity. Within the trajectory
of the dot, however, the photoreceptors are in their transient regime where the high transient
gain elicits high signals and enhances the contrast between object and background leading
to the white stripe in the image.

Piece-wise smooth optical flow estimation

A stimulus with a less complex motion boundary was applied to demonstrate the ability
of the motion segmentation chip to completely separate two different optical flow sources,
despite the low resolution of the prototype implementation. The stimulus consisted of a
sinewave plaid pattern with a linear motion discontinuity as illustrated in Figure 7.5. The

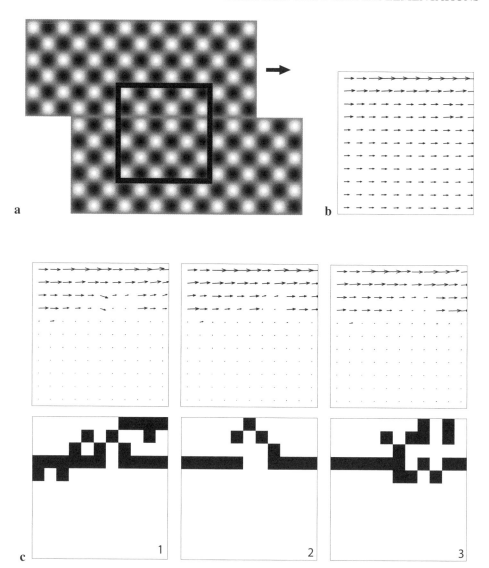

Figure 7.5 *Piece-wise smooth optical flow estimation.*
(a) The motion segmentation chip was presented with a sinusoidal plaid pattern containing
a linear motion discontinuity. (b) A single frame is shown that represents the photoreceptor
output and the optical flow signal when the discontinuity units are disabled: the optical
flow field is smooth according to the value of Bias$_{HR}$, leading to an approximately constant
and shallow vertical optical flow gradient. (c) A short sequence under the same conditions,
but with the discontinuity network enabled. The units are preferably active at the motion
discontinuity and lead to a complete separation between the two parts of the plaid pattern.
The resulting optical flow estimate preserves the motion discontinuity and is piece-wise
smooth within each separated region.

lower part of the plaid pattern remained stationary while the upper part moved horizontally with constant velocity. Spatial frequency and contrast of both parts were identical and the boundary between them was chosen such that in moments when they were in phase, the impression of an single uniform plaid pattern occurred. The visual field of the chip was approximately as indicated by the superimposed black frame. Although the motion discontinuity had a simple form, the task was very challenging because the stimulus pattern does not provide any cues for segmentation other than its underlying motion. An absolute temporal gradient measure, which would have been sufficient to segment the black dot against the white background in the previous example, would not provide enough information to find the motion discontinuity in this case.

Figure 7.5b shows the response when the discontinuity units were disabled. The estimated flow field is very smooth, spreading out far into the stationary part of the plaid pattern. The lateral smoothing was set relatively high ($\text{Bias}_{HR} = 0.8$ V) in order to obtain a piece-wise smooth flow field that does not reflect the spatial pattern of the plaid stimuli. The estimated optical flow field completely smoothed out the motion discontinuity. Clearly, a measure of the optical flow gradients alone would not be sufficient to determine where and whether at all there was a motion discontinuity present in the stimulus. Adding the difference measure of the feedback currents, however, allows the motion segmentation chip to find the true motion discontinuity despite the high smoothing conductance.[3] Figure 7.5c is a short, scanned sequence of the chip's response. It shows the optical flow estimate and the associated activity of the motion discontinuity units. The chip is now able to separate the two flow sources completely and to provide a nearly global flow estimate within each area. It finds the linear motion discontinuity with only a few discontinuity units remaining active within the moving plaid pattern, and it segments the visual scene into two areas of distinct piece-wise smooth optical flow.

7.2   Motion Selection Chip

The motion selection chip is the second extension of the smooth optical flow chip. It allows emulation of recurrent architectures such as the motion selective network proposed in Chapter 4. Unlike the motion segmentation chip, however, the motion selection chip does not contain the complete architecture on a single chip. Rather, it is a modification of the smooth optical flow chip such that it provides the necessary interfaces and programmability to establish the recurrent processing loop via an external, second processor or processing network [Stocker and Douglas 2004]. The modification consists of the addition of a 1-bit static random access memory (SRAM) cell in each unit, which permits to set the local bias conductances to either one of two globally adjustable values.

This approach, compared to the one applied for the motion segmentation chip, is more flexible in terms of implementing different computational architectures. The segregation of the implementation of the computational layers naturally follows the functional distinction between the lower level analog processing of local motion processing, which is high bandwidth, and the higher level digital selection process, which is of lower bandwidth yet can be more task- and stimulus-dependent.

---

[3] See also Figure 4.14 on page 80.

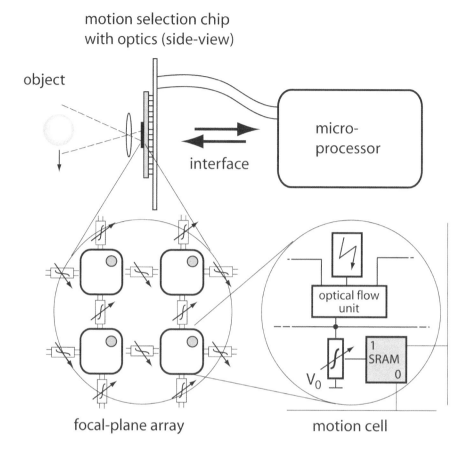

Figure 7.6 *Motion selection chip in a hybrid application framework.*
A typical hybrid system application of the motion selection chip in conjunction with an
external, digital microprocessor. The zoomed-in areas depict the cellular structure of the
optical flow network and the basic architecture of an individual unit. Each unit contains a
1-bit SRAM cell that allows control of the bias conductance. (Stocker, A. and R. Douglas,
Analog integrated 2-D optical flow sensor with programmable pixels, IEEE International
Symposium on Circuits and Systems, Volume 3, pp 9–12, © 2004 IEEE)

A typical application of the motion selection chip places it within hybrid systems as
illustrated in Figure 7.6. A sequential digital processor reads the optical flow estimate via the
on-chip scanning circuitry and a separate analog-digital converter (ADC) board, processes
this information, and sets the bias conductances of selected units of the motion selection
chip appropriately. This setup provides highest flexibility in terms of the computational
architecture of the selection process. Yet, the sequential, digital processor hardware might
be too slow or inefficient to simulate particular network architectures. The efficiency of
digital control certainly depends on the complexity of the selection task, but it also depends
on the array size of the motion selection chip. If the load is too high, however, the digital

processor could be replaced with another distributed and parallel analog processor chip that contains the appropriate selection network architecture.

A prototype of the motion selection chip has been fabricated in the same BiCMOS process as the two other presented chips.[4] The chip contains an array of 15×15 units, where each pixel is only slightly larger than in the smooth optical flow chip. The same on-chip scanning circuitry provides the read-out of the optical flow field and the photoreceptor signal. In addition, a row- and column-address register allows us to address and program the SRAM cells in individual units.

7.2.1 Pixel schematics

The circuit diagram of a single pixel of the motion selection chip (Figure 7.7) is basically identical to the optical flow unit (Figure 5.8). The difference is that each pixel contains a 1-bit SRAM cell and modified transconductance amplifiers, which together permit us to set and hold the bias conductances of individual optical flow units in one of two states. To a linear approximation, the dynamics of the optical flow estimate of each unit ij can be written as

$$\dot{u}_{ij} = -\frac{1}{C}[E_{x_{ij}}(E_{x_{ij}}u_{ij} + E_{y_{ij}}v_{ij} + E_{t_{ij}}) - \rho(u_{i+1,j} + u_{i-1,j} + u_{i,j+1} + u_{i,j-1} - 4u_{ij})$$
$$+ \sigma_{ij}(u_{ij} - u_{\mathrm{ref}})]$$
$$\dot{v}_{ij} = -\frac{1}{C}[E_{y_{ij}}(E_{x_{ij}}u_{ij} + E_{y_{ij}}v_{ij} + E_{t_{ij}}) - \rho(v_{i+1,j} + v_{i-1,j} + v_{i,j+1} + v_{i,j-1} - 4v_{ij})$$
$$+ \sigma_{ij}(v_{ij} - v_{\mathrm{ref}})], \tag{7.8}$$

where σ_{ij} can take on two values, depending on the state S_{ij} of the memory cell, thus

$$\sigma_{ij} = \begin{cases} \sigma_0 & \text{if } S_{ij} = 0 \\ \sigma_1 & \text{if } S_{ij} = 1. \end{cases} \tag{7.9}$$

The bias conductances σ_0 and σ_1 are defined by the transconductance g_m of a five-transistor transconductance amplifier. In sub-threshold operation, g_m is proportional to the bias current I_b of the amplifier. I_b is controlled by two parallel-connected bias transistors and the state of S. While one bias transistor is always on, the state of the SRAM cell directly enables or disables the second bias transistor. In this way, σ_0 is defined by the bias current set by Bias$_{Op1}$ whereas σ_1 depends on the combined bias currents set by Bias$_{Op1}$ and Bias$_{Op2}$.

These global bias voltages are freely adjustable. Typically, however, they are set so that $\sigma_1 \gg \sigma_0$ and σ_0 is small. Thus, if the memory bit is not set, the optical flow estimate is just biased strongly enough toward the reference motion vector ($u_{\mathrm{ref}}, v_{\mathrm{ref}}$) to guarantee robust behavior. On the other hand, if the memory bit is set, the large σ_1 strongly biases the local optical flow estimate and, if large enough, shunts the resistive node to the reference motion vector. Depending on the strength of σ_1 the response of a local motion processing unit can be reduced or, in the extreme, the unit can be functionally disabled.

Memory cells of individual units can be addressed and programmed individually by activating the appropriate column and row lines (*col_att* and *row_att*) in the address register,

---

[4]Exact specifications are listed in Appendix D.

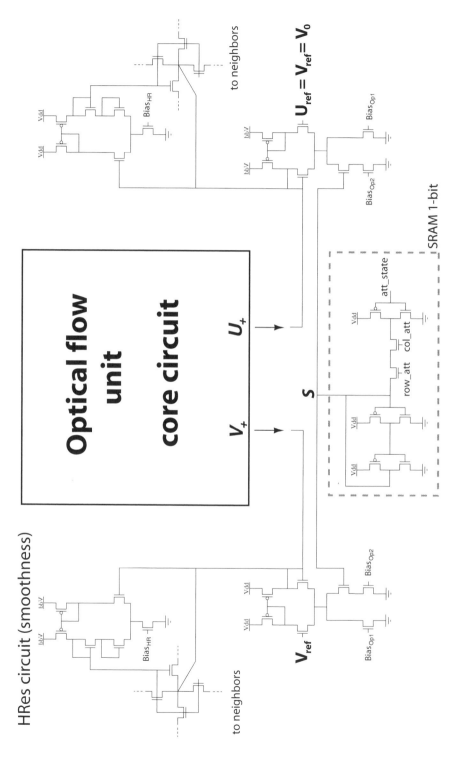

Figure 7.7 *Schematics of a single pixel of the motion selection chip.*

and setting the right bit-state on the *att_state* line. The SRAM cell continuously stores the state until a change in the local bias conductance is requested, requiring updates of only those units which change their state. The transistor sizes of the two cross-coupled inverters are designed so that the memory state S is low by default when the motion selection chip is powered up.

7.2.2   Non-linear diffusion length

The key effect of a local increase in the bias conductance is that the diffusion length in the resistive layers of the optical flow network is reduced at the particular node. The diffusion length in ohmic resistive layers is given as the root ratio $\sqrt{\rho/\sigma}$ (in the continuous approximation limit). It indicates how quickly a voltage signal attenuates along the resistive network. Thus, the diffusion length is a direct measure of the width of the region of motion integration. If the bias conductance σ is very large, integration is greatly reduced. In this way a shunted unit provides a local barrier for spatial motion integration.

In an optical flow network with ohmic smoothness conductances ρ, shunted units would significantly influence the estimates of their neighbors. This can be observed, for example, in the simulations of the motion selective system shown in Chapter 4: the flow vectors at the outer boundary of the selected areas are significantly reduced (see Figure 4.17 on page 89.). However, using HRes circuits to implement the smoothness conductances proves once more to be beneficial. Their sigmoidal current–voltage characteristics, which saturate for voltage differences larger than $\Delta V_{sat} \approx 150\,\text{mV}$, sharply reduce the diffusion length for local voltage gradients larger than ΔV_{sat}. Thus, shunted units have little effect on their neighbors when the estimated optical flow is significantly larger than the reference motion. Figure 7.8 illustrates this nicely by showing the simulated response of a simplified one-dimensional version of the motion selection chip using either ohmic or saturating lateral resistances. The dashed line in both panels represents the noisy normal optical flow estimate of the chip when smoothness is disabled. If the units are coupled and none of them are shunted (Figure 7.8a), both types of conductance lead to a smooth flow field estimate. The saturating conductances, however, induce smoothing only across small motion gradients while preserving the motion discontinuities. The difference becomes even more significant when the units seeing the lower background motion are shunted (Figure 7.8b): with saturating resistances, the optical flow estimate in the active area of the array does not differ from the non-shunted case, whereas with ohmic resistances, the low reference motion significantly affects the peak level of the optical flow estimate in the non-shunted region.

7.2.3   Experiments and results

Different experiments have been performed to illustrate the functionality of the motion selection chip and demonstrate different applications. The experimental setup was basically a digital/analog hybrid system according to Figure 7.6. The motion selective chip was interfaced to a PC via a data acquisition board that reads out the analog voltages representing the optical flow and the photoreceptor signal, and sets the row- and column-address register on the chip to address and set the state of each memory cell. Due to the relative small array size, several hundred read/write cycles per second were possible. In all of the following

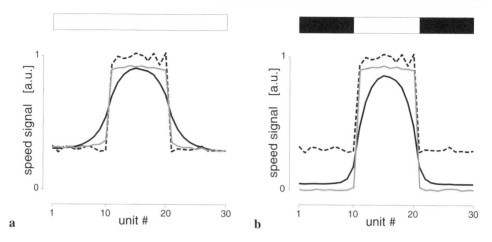

Figure 7.8 *Shunting in resistive networks.*
The optical flow network is simulated with either saturating (gray curve) or ohmic (black curve) resistances implementing the smoothness constraint. The dashed line represents the noisy normal flow estimate ($\rho = 0$) for two motion sources (object and background). The bar on top of each figure represents the state of the memory cells (black = active). (a) No shunting. (b) The units seeing the slower background motion are shunted to the reference motion $v_{ref} = 0$. (Stocker, A. and R. Douglas, Analog integrated 2-D optical flow sensor with programmable pixels, IEEE International Symposium on Circuits and Systems, Volume 3, pp 9–12, © 2004 IEEE)

experiments, the ratio of the bias conductances for activated versus non-activated shunt was in the order of 10^3.

Static selection

A first set of experiments shows the effect of statically shunting an arbitrary ensemble of units and illustrates how the non-ohmic smoothness conductances interact with the shunting of the bias conductances. As illustrated in Figure 7.9b, a static ensemble of units forming the letter "A" were shunted (disabled) in the center of the array (black = active shunts). Then, a vertically moving orthogonal sinewave plaid pattern (photoreceptor signal shown in Figure 7.9a) was presented to the chip, inducing a uniform optical flow field. Figure 7.9c shows that shunting almost completely suppresses the response in the associated units, while the effect on immediate neighbors is small due to the applied HRes circuits. Especially, it is interesting to see that the units in the small isolated active region in the center of the array were able to maintain their activity. It is important to note that the coupling between the units was enabled and that smoothness was of similar strength (Bias$_{HR}$ = 0.35 V) as in the estimated flow fields of the smooth optical flow chip shown in Figure 6.10.

A second example shows even better how the saturating resistances keep the effect of shunting localized. Instead of presenting the chip with a uniform global motion source, a split sinewave plaid pattern was applied, equivalent to the stimulus used for the motion segmentation chip in the experiment of Figure 7.5; one of the halves was moving while the other one was stationary. Figure 7.10a shows the photoreceptor response of the chip to the

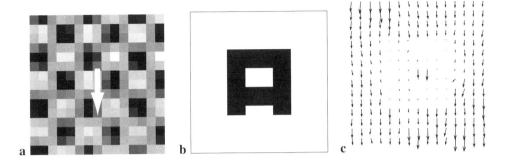

Figure 7.9 *Shunting the units of the motion selection chip.*
(a) A moving sinewave plaid pattern induces a uniform optical flow field, but the responses of units with an activated (black) shunting conductance (b) are suppressed (c). (Stocker, A. and R. Douglas, Analog integrated 2-D optical flow sensor with programmable pixels, IEEE International Symposium on Circuits and Systems, Volume 3, pp 9–12, © 2004 IEEE)

presented stimulus. Due to adaptation, the stationary pattern on the right-hand side almost completely faded out.[5] If none of the bias conductances are shunted ($Bias_{Opl} = 0.31$ V), the resulting flow field is smooth and motion information is spread into the static areas of the stimulus pattern (Figure 7.10b). However, if a small diffusion barrier is built up by shunting two columns of optical flow units at the plaid borders (Figure 7.10c), the motion discontinuity is preserved and the estimated flow field shows two distinct piece-wise uniform areas. This example demonstrates that the local control of the bias conductances in combination with the saturating lateral resistances can provide an alternative option to the control of the lateral conductances performed by the motion segmentation chip. While a two-unit boundary around distinct motion sources in an 15×15 array represents a substantial part of the total resolution, it is certainly negligible for implementations containing substantially larger arrays than this prototype chip. Controlling the bias conductances has the advantage that only one conductance per unit has to be set, which reduces the pixel size and simplifies the implementation.

Tracking by selective bottom-up attention

The last experiment illustrates how a hybrid system containing the motion selection chip can perform a selection task of the kind described in Section 4.5.2 by applying *recurrent feedback*. The goal of the system is to detect particular patterns of motion and then track these patterns by recurrently suppressing those optical flow units that do not belong to the pattern. Since the continuous dynamics (4.44) of the attention network are very inefficient to emulate on a digital sequential processor, the selection process is performed by means of a network with simplified binary dynamics. A bottom-up attention process is implemented by a hard-WTA network, which – once visual speed crosses a threshold – selects the unit of the motion selection chip that reports the highest optical flow estimate. It then shunts all other units except those that are within a given small neighborhood area around the

---

[5]Note that due to adaptation, the chip treats the static pattern basically as a uniform background. See also the discussion on adaptation and the robustness of the brightness constancy constraint in Section 5.2.2.

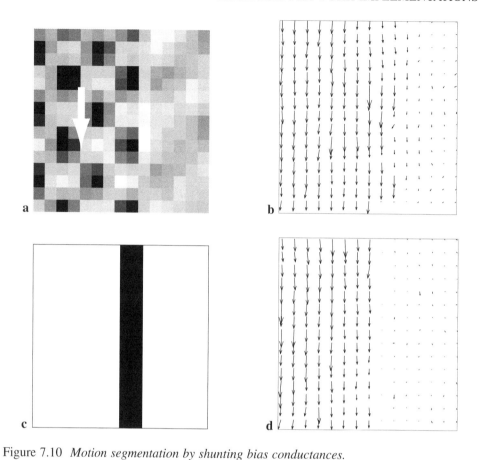

Figure 7.10 *Motion segmentation by shunting bias conductances.*
(a) The photoreceptor signal for a split plaid pattern where one half was moving while the other half was stationary. Note that adaptation leads to the much reduced photoreceptor signal in the stationary part on the right. (b) Without shunting, the estimated flow field smooths out the motion discontinuity. (c) Applying a two-unit-wide shunting barrier along the plaid boundaries leads to (d), a clearly separated, piece-wise smooth optical flow estimate.

winning unit, by setting their bias conductances high. Figure 7.11 shows the behavior of the hybrid system for such a binary selection process when presented with three motion sources (two moving black dots on a stationary white background). The panels show four sequential frames of the output of the motion selection chip, containing the photoreceptor signal, the stored SRAM bit-states, and the estimated optical flow. The first frame shows the starting condition before the WTA competition has been activated. All bias conductances are low and the sensor provides a smooth optical flow estimate for both objects. As soon as the recurrent loop is activated, the WTA network selects a location on one of the moving objects and shunts all units outside a predefined neighborhood area around the selected winning unit, thus suppressing the optical flow estimate of these units. Here, the selection of the dot on the top in the shown example is random because both dots are actually moving

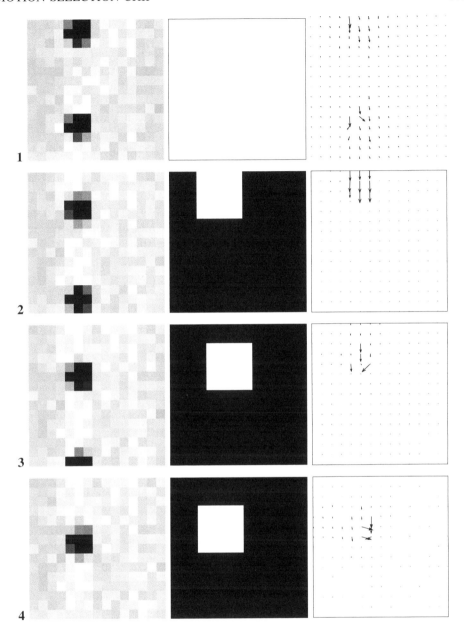

Figure 7.11 *Tracking by WTA competition and recurrent feedback.*
Black dots on white background moving down. A WTA selection network recurrently
controls the bias conductances of the motion selection chip. It shunts the bias conductances
of all units in the array that do not belong to the neighborhood around the unit with highest
optical flow estimate. Frame #1 shows the states right before the feedback loop is activated.
Columns from left to right: photoreceptor output, conductance states (low = white, high =
black), and optical flow estimate. (Stocker, A. and R. Douglas, Analog integrated 2-D
optical flow sensor with programmable pixels, IEEE International Symposium on Circuits
and Systems, Volume 3, pp 9–12, © 2004 IEEE)

with equal velocity. However, once an object is selected the recurrent shunting guarantees that the next winning unit can only emerge from the subset of active units. In this way, the selected object is tracked by dynamically adapting the shunting pattern. Once the object leaves the visual field or stops moving, threshold is not met and shunting is released in the whole array, allowing any new moving object entering the visual field to win the WTA competition.

Note that the shunting conductances were set rather high in all the above examples ($Bias_{Op2} = 0.7$ V), which leads to the complete suppression of the optical flow estimate outside the selected areas. This is not necessary because the strength of the shunting conductance is freely adjustable. A lower bias setting would lead to only a partial weakening of the optical flow estimate, eventually permitting an object outside the selected area to win the competition and gain the attentional spot-light, if it were to move significantly faster than the currently selected object. Such partial suppression would implement a classical scheme of attentionally modulated perception, where the signal strength outside the focus of attention is relatively week compared to the signals from the attended area.

8

Comparison to Human Motion Vision

This book began with a description of the demanding task of a football goal-keeper estimating the relative motion between the ball and himself. Yet, aside from this initial illustration of the problem, the development of the optical flow network architecture, its network extensions for motion segmentation and selective attention, and their aVLSI implementations were purely guided by a rational perspective on what properties a visual motion estimation system has to have in order to solve the task. At no moment was there a motivation to imitate any existing biological solution as was the goal in other approaches.[1] Considering the large differences in the computational substrate between silicon and nervous tissue, a close structural embodiment of biological solutions does not seem necessarily reasonable.

However, the rationale and logic that guided the development of the presented networks is human, and one might wonder how far they are influenced by the human perspective on the visual world, that is by the human visual motion system. This chapter will draw a comparison between the presented silicon visual motion systems (specifically the smooth optical flow chip) and the human visual motion system. It will compare the perceptual behavior of human subjects and the chip in some typical psychophysical experiments and loosely discuss similarities in the computational structure of both systems.

8.1 Human vs. Chip Perception

The human visual motion system has been well studied and extensively described in the psychophysical literature. One reason why this is such an interesting research area is because humans do *not* perceive visual motion veridically. Rather they misjudge the motion of a

---

[1]Prominent representatives are the circuits that emulate the fly's visual motion system; see the review on aVLSI motion circuits in Section 2.7.

moving visual stimulus depending on many other stimulus properties. Furthermore, human visual motion perception is strongly influenced by contextual information and attention as shown by early *Gestalt psychology* [Wallach 1935]. Consequently, most psychophysical experiments have been limited to rather simple well-defined stimuli such as drifting gratings and plaids, with a limited number of stimulus parameters. In the following, a few of these experiments have been selected and replicated for the smooth optical flow chip, and the result are compared to data from human subjects.

8.1.1   Contrast-dependent speed perception

Human perception of visual motion is inversely dependent on stimulus contrast: a low-contrast stimulus is perceived to move slower than a high-contrast one [Thompson 1982, Stone and Thompson 1992]. The reader can immediately and easily verify at least the extreme case of this effect: when closing the eyes, thus reducing the contrast to zero, there is typically no perception of visual motion. As demonstrated in Chapter 4, an inverse contrast dependence of perceived visual motion results from including a bias constraint in the optical flow model. Although this contrast dependence might appear at first glance to be a shortcoming, it can be viewed as optimal given that the chosen reference motion $\vec{v}_{\mathrm{ref}}$ and the form of the bias conductances σ are in agreement with the statistical distributions of the experienced visual motion of the observer [Simoncelli 1993, Weiss et al. 2002, Stocker and Simoncelli 2005a]. The work of Simoncelli and colleagues showed that a bias toward zero motion can account quite well for the observed contrast dependence in human visual motion perception. The smooth optical flow chip also has a preference for zero motion since $\vec{v}_{\mathrm{ref}} = \vec{0}$ ($U_{\mathrm{ref}} = V_{\mathrm{ref}} = V_0$; see Figure 5.8).

Contrast-dependent visual speed perception of human subjects and of the smooth optical flow chip was compared. Humans were asked in a psychophysical experiment to compare the speed of two drifting sinewave gratings. One grating (reference) had fixed speed and contrast while the second grating (test) was of varying contrasts and speeds. Otherwise, the two gratings were identical. Using a typical two-alternative-forced-choice (2AFC) protocol, the speed ratio that was necessary to make both gratings appear to move equally fast was determined. Typical results for a subject human are shown in Figure 8.1 (data replotted from [Stocker and Simoncelli 2005a]). The data points represent the relative speed of the test stimulus at different contrast levels for a given reference stimulus with fixed contrast and speed. Clearly, for low contrasts, the test grating has to move substantially faster in order to be perceived at the same speed. At contrasts equal or close to the reference contrast, the effect diminishes.

For comparison, the response of the smooth optical flow chip in the same experiment is also plotted in Figure 8.1. The contrast dependence of its visual speed estimate is similar to that of a human. Differences appear as a lower absolute sensitivity on contrast and a higher gain in the dependency at low contrasts. Matching in contrast sensitivity cannot be expected because the photoreceptor array is by far not as sensitive as the human retina. The difference in shape of the two curves reflects the different encoding of contrast in the two visual motion systems: while contrast is rather linearly represented in the chip as the spatial gradient of brightness, its effect upon the neural behavior of the visual

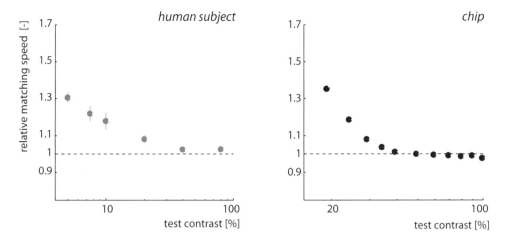

Figure 8.1 *Contrast-dependent visual speed perception: Human vs. chip.*
The relative speed of a test grating necessary to match the perceived speed of a reference
grating with fixed speed (1 deg/s) and contrast ($c = 50\%$) is plotted as a function of test
contrast. At test contrasts $<50\%$, the test grating had to move faster in order to match the
reference stimulus. Human data are replotted from [Stocker and Simoncelli 2005a].

cortex seems to follow a compressive non-linearity (e.g. sigmoidal function), with an early
saturation point [Sclar et al. 1990, Hürlimann et al. 2002, Stocker and Simoncelli 2005a].
Qualitatively, however, both systems behave quite similarly.

8.1.2   Bias on perceived direction of motion

Stimulus properties such as contrast not only have an influence on the human perception
of speed but also lead to a biased perception of the direction of motion, away from the
true retinal stimulus motion (e.g. [Stone et al. 1990]). Such bias was investigated in the
smooth optical flow chip's response to various moving plaid patterns. Plaids are a class
of two-dimensional visual stimuli that are generated by the superposition of two one-
dimensional sinewave gratings of different orientation. Typically, plaids are differentiated
into *type-1* and *type-2* plaids and an example of each type is shown in Figure 8.2. Plaids
for which the direction of pattern motion lies in between the directions of motion of
the individual gratings are of type-1, whereas if the pattern motion direction lies outside
of the component motion direction they are of type-2. Plaids have been used for many
physiological and psychophysical experiments because they represent well-constrained and
simple two-dimensional motion stimuli.

   The response of the smooth optical flow chip was recorded as a function of individual grating contrast for a type-1 and a type-2 drifting plaid pattern. Figure 8.2 illustrates
the composition of the two plaid patterns and shows the measured global motion estimates of the chip, recorded for different contrast ratios of the two gratings in each plaid.

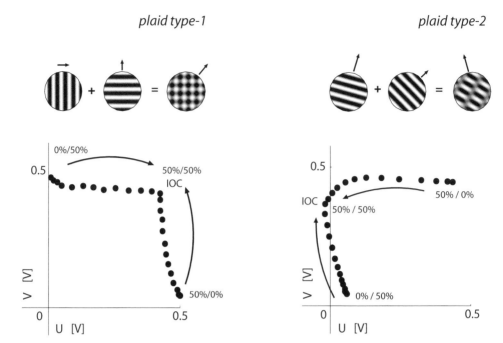

Figure 8.2 *Type-1 and type-2 plaid responses of the smooth optical flow chip as a function of underlying grating contrasts.*

Each data point represents the time-averaged global motion estimate. As the contrast of one of the plaid gratings is decreased, the estimated direction of motion shifts from the intersection-of-constraints (IOC) solution toward the component motion direction of the higher contrast grating.

The type-1 plaid consisted of the superposition of two orthogonal gratings with the same spatial and temporal frequency. When both gratings had the same contrast [50%/50%], the response was in good agreement with the intersection-of-constraints (IOC) solution, which is equivalent to the vector-average (VA) solution for this particular plaid pattern. As the contrast of either one of the gratings was decreased while the other remained at 50%, the chip's reported motion continuously changed direction until it represented pure component motion of the high-contrast grating. The same procedure was applied for the type-2 plaid stimulus. The gratings had the same spatial frequency yet their temporal frequencies differed such that the pattern motion (IOC) was clearly away from the mean of the component motions (VA). When the contrast of both gratings was equally high [50%/50%], the chip reported the correct pattern motion. Again, as the contrast of either of the two gratings was continuously decreased, the output increasingly shifted toward the individual component motions.

The behavior of the optical flow chip in the two plaid experiments is qualitatively in good agreement with human psychophysical data. It has been shown, for example, that humans perform an IOC computation for type-1 and type-2 plaid patterns if the contrast ratio of the involved gratings is not too large and spatial frequencies are similar [Adelson and Movshon 1982]. Furthermore, human subjects report increasing deviations of the perceived pattern motion from the IOC solution toward the normal motion of the higher contrast grating, if the contrast of the other grating is successively reduced. This matches the continuous transition from the perception of the VA to the IOC motion direction, depending on the contrast differences of the gratings [Stone et al. 1990].

Human perception of motion of a type-2 plaid pattern depends not only on the relative contrast but also on the separation angle between the underlying gratings. The perceived direction of motion for increasing angles thereby changes from the VA to the IOC direction. In order to compare human with chip perception, the psychophysical experiment of Burke and Wenderoth [1993] was replicated for the smooth optical flow chip. The time-averaged response was measured for four different separation angles, where the temporal frequencies of the gratings were modified such that the pattern motion of the plaid was constant under all conditions. Figure 8.3 compares the directional bias of human and chip perception. Both systems behave qualitatively similarly, asymptotically approaching the IOC solution with increasing separation angle. Compared to the human subject, the chip shows saturation at narrow angles and its estimate approaches the IOC solution only at very wide separation angles. This is due to the relative strong weight of the bias constraint in this experiment: the strength of the bias conductance had to be slightly increased ($\text{Bias}_{OP} = 0.34$ V) compared to previous measurements (see e.g. Section 6.1) in order to assure a robust behavior under the given experimental conditions, in particular because of the chip's sensitivity to the refresh rate of the electronic stimulus generator.

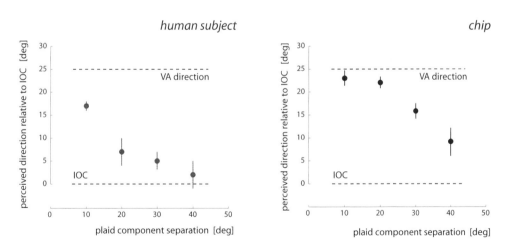

Figure 8.3 *Perceived directional bias from true pattern motion of a type-2 plaid with different separation angles.*
The separation angle of a type-2 plaid was varied while the temporal frequencies of the underlying gratings were modified such that the pattern motion (IOC) was constant. With increasing separation angle, the perceived direction of motion of both the chip and a human subject shifts in a qualitatively similar way from the VA to the IOC direction (dashed lines). Human data and experimental setup were adapted from [Burke and Wenderoth 1993].

8.1.3 Perceptual dynamics

The above experiments are just a small subset to illustrate the similarities between human visual motion perception and the smooth optical flow chip's response. In fact, more psychophysical experiments can be fitted with an optimal estimation model that uses the brightness constancy constraint and a bias constraint similar to the one incorporated in the smooth optical flow chip [Weiss et al. 2002]. These optimal models, however, only account for the steady-state responses that have been considered and compared so far. Yet both humans and the smooth optical flow chip are dynamic systems. It is interesting to see if the similarities also extend to their perceptual dynamics.

There have been psychophysical experiments to investigate the dynamics of human visual motion perception. When presented with a type-2 plaid, for example, humans perceive visual motion to be biased toward the VA direction of the component motions at short durations, approaching the IOC solution only after some time lag [Yo and Wilson 1992]. This particular experiment has been replicated for the chip, and the results are shown in Figure 8.4, together with the original human data. Surprisingly, the dynamics of both systems seem qualitatively very similar. At short time-scales the percept is biased toward the VA motion direction, and only later does it converge asymptotically toward the IOC solution. In both systems, the time constant increases with decreasing total plaid contrast (not shown), with a simultaneous increase in the offset of the asymptotic estimate of motion direction from the IOC solution. The asymptotic estimate represents the steady-state percept as reported in the previous experiments. Even a very long integration time does not allow the systems to converge to the true pattern motion direction at low contrast values.

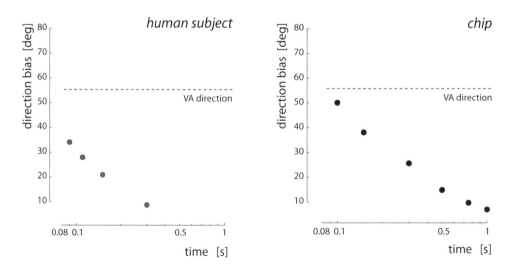

Figure 8.4 *Perceptual dynamics of humans and the smooth optical flow chip.*
The figure shows the directional bias of a human subject and the chip as a function of presentation time. The experimental setup and the replotted data of the human subject data are from [Yo and Wilson 1992].

There are not many other experiments reported in the literature that investigate the dynamics of human visual motion perception. The results of Yo and Wilson's experiment, however, suggest that the dynamics of human visual motion perception resemble the steepest gradient dynamics of the smooth optical flow chip. Hence it appears that the human visual motion system not only employs an optimal estimation strategy with a bias for slow speeds, but also tries to approach the optimal solution along the shortest path possible in perceptual space.

8.2   Computational Architecture

Since the perceptual behavior of the smooth optical flow chip seems surprisingly similar to human visual motion perception, the question arises whether this is reflected in some strong structural similarities between both systems. We know a fair amount about the physiological processes underlying human visual motion perception. For obvious reasons, however, our physiological knowledge of the human brain is largely limited to non-invasive analytical methods. These methods, such as functional magnetic resonance imaging (fMRI), do not yet offer the spatial and temporal resolution necessary to extract the fine structural and functional organization of the human brain, although technological advances are happening at a fast pace. Fortunately, it seems that the organization and physiological processes in the early visual cortex of primates are very similar amongst different species. Thus, the following short physiological description is mainly based on data from macaque monkeys, but is assumed to be a good model for the human visual cortex.

Processing of motion information in primates' visual cortex is assumed to involve two stages, as illustrated in Figure 8.5. The primary visual cortex (V1) serves as the first stage. Neurons in area V1 are sensitive to the orientation and retinal position of visual stimuli. This is true for monkeys and cats [Hubel and Wiesel 1962]. Many of these neurons also have space–time-oriented receptive fields, and so respond preferentially to a stimulus moving in a particular direction. The small receptive field sizes of V1 neurons, however, inherently generate an aperture problem. These neurons cannot estimate object motion (or *pattern motion*) even if they had the computational means for it. They can only encode *component motion* of a visual stimulus, which is the motion orthogonal to the local brightness gradient. Also, their separable and narrow spatio-temporal frequency tuning does not allow them to explicitly represent local velocities. A second processing stage is required that (i) resolves the aperture problem due to the small receptive fields by spatial integration of visual motion information and (ii) locally integrates the responses of those V1 neurons with a spatio-temporal frequency tuning that is consistent with a particular local motion vector (speed and direction). It is widely believed that area MT is the earliest stage where subpopulations of neurons show the above properties [Adelson and Movshon 1982, Movshon et al. 1985].

A substantial fraction of MT neurons exhibits an explicit velocity encoding scheme [Perrone and Thiele 2001, Priebe et al. 2003, Anderson et al. 2003, Priebe and Lisberger 2004]. Several models of velocity encoding in MT have been proposed in which neurons receive the combined input from spatio-temporally tuned V1 neurons that are consistent with a particular local motion. These neurons populate a plane in the spatio-temporal frequency domain of visual input where each plane corresponds to a particular local optical flow vector [Heeger et al. 1996, Simoncelli and Heeger 1998]. It has been argued that this integration process is not stimulus-dependent, and thus includes an unbiased sample of V1 neurons that span the complete range of spatial orientations [Schrater et al. 2000].

The V1 neurons that project to area MT are predominantly *complex cells*, a particular prevalent sub-class of neurons in V1 [Movshon and Newsome 1996]. Complex cells show similar direction, orientation, and spatio-temporal frequency tuning like *simple cells*, which are another large sub-class of V1 neurons. Unlike simple cells, however, their responses are relatively invariant to the exact location of the stimuli within their receptive field. These properties led to the notion that responses of V1 complex cells approximately encode the local motion energy of the stimulus, computed as the sum of squared responses of even- and odd-symmetric oriented linear filters [Adelson and Bergen 1985, Heeger 1987a, Grzywacz and Yuille 1990]. Clearly, local visual ambiguities (*e.g.* the aperture problem) and noise leads to motion energy components in multiple planes of the spatio-temporal frequency input space. As a consequence, several velocity-tuned neurons in MT will be active, and there has to be a competitive process taking place in order to generate an unambiguous vote for a particular local velocity. Although neurons in area MT have rather large receptive fields, they encode primarily local translational visual motion information, similar to an optical flow field. More global and complex optical flow patterns such as those radial and rotational flow fields originating from ego-motion are discriminated by neurons in area

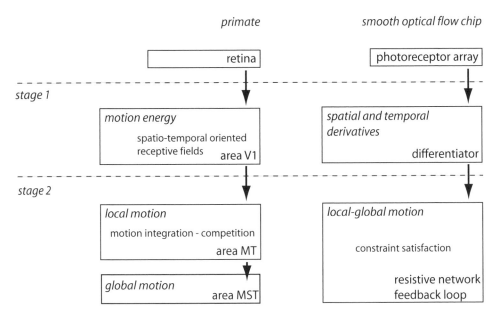

Figure 8.5 *Simplified computational structure of primate and silicon motion vision.*
Both systems compute visual motion in a two-stage process. While the first stage essentially applies spatio-temporal filters to the visual input, the second stage is required to combine these filter outputs appropriately, and to apply a competitive process to find a unique estimate of local visual motion.

MST, an extra-striate area that receives its input dominantly from area MT [Duffy and Wurtz 1991a, Duffy and Wurtz 1995].

The *smooth optical flow chip*, the simplest of the circuits presented, exhibits a somewhat similar two-stage functional organization. After visual information is transduced by a photoreceptor array, spatial and temporal gradients of the logarithmically compressed and adapted brightness signals are computed. These gradients are input to the second stage, the recurrent two-layer network that solves the constraint satisfaction problem by minimizing the cost function (4.19). Here, visual motion is integrated and the network converges to a solution that is most consistent with the local observation (brightness constancy constraint), the motion in its surroundings (smoothness constraint), and prior knowledge about the visual environment (bias constraint).

Given the similarities in perceptual behavior and computational structure between the human visual motion system and the smooth optical flow chip, it seems natural to assume that both systems apply similar computations to converge to similar solutions for the problem of visual motion estimation. Thus, I will propose and discuss in more detail possible interpretations of the neural responses along the cortical motion pathway with respect to the constraint optimization problem (4.6). Of course, since there are still many unresolved questions in understanding the neural mechanisms of visual motion processing, these interpretations are rather speculative, yet propose hypotheses that may be tested in physiological experiments.

First, assume that neurons in area MT encode the perceived visual motion. This assumption is supported by studies that show strong correlations between neural responses in area MT and the perception of visual motion [Newsome et al. 1989, Britten et al. 1996, Shadlen et al. 1996]. Figure 8.6 illustrates schematically an idealized population of MT neurons responding to a moving visual stimulus.[2] Each neuron has an identical tuning curve centered around its individual, preferred speed. The neuron whose preferred speed is closest to the perceived speed responds the most. In comparison, the response of the optical flow chip represents the point of minimal cost. The two representations seem inversely related to each other. Thus, the response of a neuron with given preferred motion, $\vec{v}_i$, corresponds to a particular value $H(\vec{v}_i)$ of the cost function (4.19). With this relationship in mind, it is possible to relate each of the three constraint terms in the cost function to MT neural responses in the following ways:

- The *brightness constancy constraint* term $F(\vec{v}_i) = \left(E_x u_i + E_y v_i + E_t\right)^2$ is essentially minimal for the MT neuron whose preferred motion best matches the stimulus motion. The expression can be expanded, leading to a sum of terms that are products of spatio-temporal brightness gradients. These terms can be interpreted as typical responses of V1 neurons projecting to the particular MT neuron [Simoncelli 2003]. A direct mapping from the brightness constancy constraint to neural responses via an inverse activation function, however, does not seem plausible because it requires afferent synaptic input to be minimal for the most and maximal for the least excited MT neuron. Especially for stationary visual stimuli, such an inhibition-dominated mapping does not appear to match the physiology. It seems more plausible that V1 neurons whose spatio-temporal tuning is consistent with a particular local motion form

---

[2]For simplicity, the problem is reduced to the one-dimensional case; hence, the local motion vector (optical flow vector) reduces to a scalar value (speed).

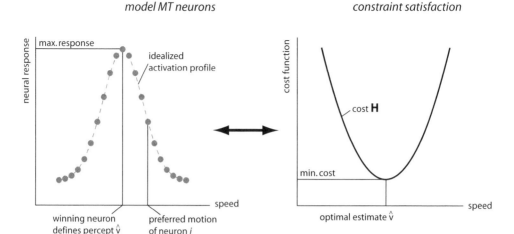

Figure 8.6 *Neural encoding and constraint satisfaction.*
Visual motion is possibly encoded in the population activity of MT neurons where the preferred motion of the winning neuron (or eventually the mean of the population) represents the estimate. For the smooth optical flow chip, the optimal estimate is at the lowest point on the cost function (4.19). Considering the neural population activity profile as an approximate inverse of the cost function generates a computational link between neural responses and the constraint satisfaction problem.

excitatory projections to a common MT neuron, while inhibiting other MT cells [Koch et al. 1989, Simoncelli and Heeger 1998]. In this way, the synaptic input from V1 is maximal for the MT neuron with preferred motion closest to the stimulus motion.

- The *smoothness constraint* can be mapped by assuming local excitation of retinotopically close MT neurons tuned for similar local motions. The existence of such excitatory connections in MT is not yet certain, although similar patterns of excitatory connections have been reported in lower visual areas such as area V1 [Rockland and Lund 1983].

- The *bias constraint* is minimal if the motion estimate is closest to a particular reference motion. For humans this reference motion seems to be zero motion as suggested by the previously shown psychophysical results. The bias constraint could be reflected in different ways in the responses of MT neurons. A first possibility is that each MT neuron receives a constant excitatory input that is the larger, the closer the neuron's preferred motion is to the reference motion. The effect is that for weak motion signals at low stimulus contrast this constant synaptic input dominates and induces the shift in perceived motion. This would be a direct implementation of the bias constraint, which, however, seems metabolically not very efficient.

  A second way to understand the bias constraint in a neural sense is to assume different response characteristics for MT neurons tuned for different speeds. For example, the contrast response function $r_i(c)$ of MT neurons can be described as a typical sigmoidal function

$$r_i(c) = \alpha \frac{c^n}{c^n + c_{50}^n} + \gamma \tag{8.1}$$

where α and γ are constants [Sclar et al. 1990]. Assume that neurons tuned for low speeds saturate earlier (lower c_{50} value) and/or have higher gain (higher exponent n) than neurons tuned for higher speeds. Then, reducing the stimulus contrast reduces the responses of low-speed neurons *less* than those of high-speed neurons, which leads to a shift in the position of the winning neuron toward lower speeds. This is illustrated in Figure 8.7. The exact form of the bias depends on the parameters of the contrast response function and how they change with preferred speed. We did some preliminary studies to find physiological evidence for such dependencies, yet the available data do not sufficiently constrain this possibility so far [Stocker and Simoncelli 2005b].

And finally, changes in speed tuning represent another possible neural mechanism of implementing the bias constraint. The idea is that the speed-tuning curves of MT neurons change with stimulus parameters such as contrast in a way that is consistent with the shift in perceived visual motion. Physiological evidence for contrast-induced shifts in tuning curves, however, seems controversial: some studies that used grating stimuli do not see a shift in preferred speeds [Priebe et al. 2003], whereas others, using random dot stimuli, report clear shifts [Pack et al. 2005].

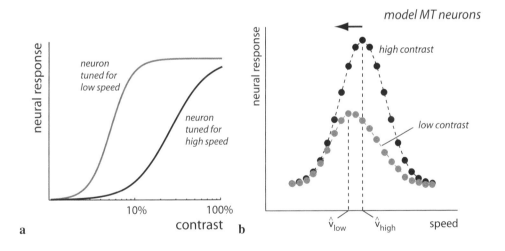

Figure 8.7 *Neural implementation of the bias constraint.*
(a) Different response characteristics of MT neurons tuned for different visual speeds can implement the bias constraint. For example, if neurons tuned for high visual speed have a contrast response characteristics that saturates later and/or has higher bandwidth (less gain), then reducing contrast leads to a shift in position of the winning neuron toward lower speeds (b).

So far, known physiological data do not promote any particular one of the above implementation options. Only further experiments and analyses will help to constrain a unique mechanism.

A neural model based on the above mechanisms assigns maximal excitation to the MT neuron whose preferred motion is closest to the desired optimal estimate. For a robust neural implementation, normalization seems necessary to provide clear MT responses over a wide range of motion energy amplitudes. MT population responses should be strong enough even for low overall amplitudes in order to allow a selection of a winning neuron. A competitive soft-WTA network as outlined in Chapter 3 could perform such normalization; it amplifies the population response profile and strengthens the response of the winning neuron, leading to a robust estimate of local visual motion. The soft-WTA network relies on local excitation and global inhibition. Interestingly, it has been shown that inhibition is crucial in generating the direction and speed tuning of MT neurons [Mikami et al. 1986]. Response strengths are significantly modulated by inhibitory effects from neurons with different speed and direction tuning [Pack et al. 2005].

As shown, not only is the steady-state visual motion percept of humans similar to the optimal solution of the constraint satisfaction problem, but also the temporal evolution of the percept resembles the gradient descent dynamics of the smooth optical flow chip (Figure 8.4). Assuming that neurons in area MT represent the local visual motion estimate, such dynamics should be reflected in the dynamics of the population activity in MT. In fact, results from recent physiological experiments support this [Pack and Born 2001, Smith et al. 2005]. In these experiments, the direction tuning curves of MT pattern cells were measured at different times after stimulus onset. MT pattern neurons can locally approximate the IOC solution. Presented with stimuli where the pattern motion was different from the component motions, the direction tuning curves changed over time. In the early response phase after stimulus onset, the direction tuning curves were strongly biased toward the vector average of the component motions. Only after a time delay of 60 milliseconds did the direction tuning become consistent with the neuron's preferred direction of the overall pattern motion. This behavior matches exactly the gradient descent dynamics of the optical flow network: the gradient of the cost function is steepest in the direction of highest local brightness contrast, which is in the direction of the component motions.

8.3   Remarks

The similarities between human visual motion perception and the response behavior of the optical flow network, and its implementation, are surprising, especially since there was no intention to copy biology. Under the assumption that the human visual system provides an efficient solution refined through millions years of evolution, these similarities indicate two things.

First, it seems that the chosen model for local visual motion estimation is a good choice, at least when considering applications on autonomous systems that face similar motion perception tasks as human beings. The chosen constraints seem to capture the essential properties of visual motion estimation, and the formulation as an unconstrained optimization problem represents a sensible way of combining the constraints in a "soft" way in order to constrain the solution. The bias constraint plays a particular and important role. Introduced as a reasonable way of keeping the computational problem well-defined, it improves the quality of the estimate by leading to statistically optimal solutions: although it seems rather counter-intuitive, a biased estimate can represent the optimal estimate as shown earlier. Gradient descent seems to be an efficient way of finding the optimal estimate of the constraint satisfaction problem. Interestingly, at least in a local sense, gradient descent is the optimal strategy with respect to finding the most direct path toward the optimal solution in "motion space". There exist other algorithmic search strategies to find the global optimum. Yet these are typically not continuous and local, and cannot be implemented in a modular computational structure, whereas gradient descent "naturally" reflects the dynamics of the analog optical flow networks.

Second, because of the similarities, the optical flow network represents a good *static and dynamic model* for human visual speed perception, although on a higher abstraction level than other models [Simoncelli and Heeger 1998]. Thus, it permits us to investigate how such optimal estimation of visual motion is physiologically implemented. We have just begun to address this question [Stocker and Simoncelli 2005b] and I have outlined some of the ideas above. Clearly, the apparent similarity between the perceptual dynamics of humans and the gradient descent dynamics of the optical flow network provides further constraints that might allow us to identify the appropriate neural mechanisms, or to make predictions that can be physiologically tested.

Of course, the similarities between the human visual motion system and the smooth optical flow chip are restricted to only local and translational motion perception. Human motion vision is much more sophisticated than just computing local motion. For example, humans can deal with transparent motion, motion in depth, rotational, and radial motion. Human visual motion perception is also modulated by attention and other higher-level cognitive top-down processes. Nevertheless, it is encouraging that at least on this simple level, the characteristics of the derived optical flow network and its implementation in aVLSI are supported and justified by behavioral similarities with the highly evolved human visual motion system.

Appendix A

Variational Calculus

A.1 Optimization Problem

Consider the following minimization problem. Given the *Lagrange function* $L : \mathbb{R}^m \times \mathbb{R}^m \times \mathbb{R}^n \to \mathbb{R}$ which is sufficiently regular (*e.g.* $\in C^2$) and the function $f : \partial\Omega \to \mathbb{R}^m$, find function $q : \Omega \to \mathbb{R}^m$, $\Omega \subset \mathbb{R}^n$ such that

$$H = \int_\Omega L(q(\vec{x}), q'(\vec{x}), \vec{x}) \ d\vec{x} \tag{A.1}$$

is minimal and $\quad q(\vec{x}) = f(\vec{x}), \quad \vec{x} \in \partial\Omega \quad$ (boundary condition).

Definition 1. A function $q(\vec{x})$ is called a candidate for a solution of (A.1) if it is defined on Ω, obeys the boundary condition, and is piece-wise differentiable.

A necessary condition for a candidate $q_0(\vec{x}) = (q_{0_1}, \dots, q_{0_m})$, $\vec{x} = (x_1, \dots, x_n)$ to be a solution of (A.1) is that the following system of differential equations applies:

$$\sum_{i=1}^n \frac{\partial}{\partial x_i} L_{\frac{\partial q_j}{\partial x_i}}(q, q', \vec{x}) - L_{q_j}(q, q', \vec{x}) = 0 \qquad j = 1, \dots, m. \tag{A.2}$$

This fundamental theorem was proven by Leonhard Euler in 1744. With this proof, he created the mathematical field of variational calculus.

A.2 Well-posed

Definition 2. An analytical problem is said to be *well-posed* according to Hadamard [Tikhonov and Arsenin 1977] if its solution (i) exists, (ii) is unique, and (iii) depends continuously on the data. Otherwise, it is said to be *ill-posed*.

Jacques Hadamard provided the following very insightful quote addressing the relation between well-posedness and physical systems:

Analog VLSI Circuits for the Perception of Visual Motion A. A. Stocker
© 2006 John Wiley & Sons, Ltd

> *But it is remarkable, on the other hand, that a sure guide is found in physical*
> *interpretation: an analytic problem always being correctly set ('well-posed'), in*
> *our use of the phrase, when it is the translation of some mechanical or physi-*
> *cal question.* (Jacques Hadamard in: *Lectures on Cauchy's Problem in Linear*
> *Partial Differential Equations* Yale University Press, 1923)

Side note:

The significance of Hadamard's quote became apparent in an early phase of my graduate studies. Working on the aVLSI implementation of the Horn and Schunck [Horn and Schunck 1981] optical flow algorithm, I realized that it could never be precisely implemented in a chip because of the finite output conductances of the feedback nodes (see Figure 4.2, page 56). As shown, the Horn and Schunck algorithm is not well-posed for every visual input. However, a physical system does not tolerate an ill-posed state. Thus "naturally", the physical system makes sure that it is always in some defined state by not providing ideal current sources. Of course, the finite output conductance imposes deviations from the original Horn and Schunck algorithm. Later, I realized the benefits of such an additional constraint – leading to the formulation of the bias constraint.

A.3   Convexity

Convex sets

Definition 3. The set $M \subset \mathbb{R}^n$ is said to be *convex* if all points on the connecting line between $u_1, u_2 \in M$ are in M. Hence, M is convex if for all $u_1, u_2 \in M$ and $t \in [0, 1]$

$$u_1 + t(u_2 - u_1) = tu_2 + (1 - t)u_1 \quad \in M. \tag{A.3}$$

Examples of convex sets are the space $\mathbb{R}^n$ and intervals in $\mathbb{R}^n$.

Convex functions

Definition 4. A function $f(u)$ defined on a convex set M is called a convex function if

$$f(tu_2 + (1 - t)u_1) \le tf(u_2) + (1 - t)f(u_1) \quad \forall(u_1, u_2) \in M \text{ and } \forall t \in [0, 1]. \tag{A.4}$$

In particular, it is said to be *strictly convex* for $u_1 \ne u_2$ if

$$f(tu_2 + (1 - t)u_1) < tf(u_2) + (1 - t)f(u_1) \quad \text{for } 0 < t < 1. \tag{A.5}$$

Lemma 1. For a continuous differentiable strictly convex function $f(u)$ defined on a convex set M and $(u_1, u_2) \in M$ the following inequality holds:

$$f(u_1) + \operatorname{grad} f(u_1)(u_2 - u_1) < f(u_2) \quad \text{for } u_2 \ne u_1. \tag{A.6}$$

Proof. According to the Taylor expansion $f(u)$ can be rewritten for every $(u_1, u_2) \in M$ as

$$f(u_1 + t(u_2 - u_1)) = f(u_1) + \operatorname{grad} f(\hat{u}) t(u_2 - u_2) \quad \text{with } t \in [0, 1], \tag{A.7}$$

where $\hat{u}$ is between u_1 and u_2. According to the definition of a strictly convex function and recognizing that the left-hand sides of (A.5) and (A.7) are equivalent, the inequality

$$f(u_1) + \operatorname{grad} f(\hat{u}) \, t(u_2 - u_2) < tf(u_2) + (1 - t)f(u_1)$$

holds. It can be simplified (division by t, $0 < t < 1$ (A.5)) such that

$$f(u_1) + \operatorname{grad} f(\hat{u})(u_2 - u_1) < f(u_2).$$

Since $f'(u)$ is continuous, $\lim_{\hat{u} \to u_1}$ finally leads to

$$f(u_1) + \operatorname{grad} f(u_1)(u_2 - u_1) < f(u_2). \tag{A.8}$$

This holds for all pairs (u_1, u_2) which are allowed in (A.5) and thus the proof is complete. It follows directly from Lemma 1 that u_1 is a global minimum of $f(u)$, if $\operatorname{grad} f(u_1) = 0$. $\qquad \square$

The Hessian matrix

Definition 5. The Hessian $\mathbf{J}$ of a twice continuously differentiable function $f(x)$ with $x = (x_1, \ldots, x_n)$ is a matrix where the entry at each position i, j is defined as

$$\mathbf{J}_{i,j} = \begin{cases} \partial^2 f(x)/\partial x_i^2 & \text{for } i = j \\ \partial^2 f(x)/\partial x_i \partial x_j & \text{otherwise.} \end{cases} \tag{A.9}$$

Lemma 2. (with no proof) A function $f(x)$ is convex if its Hessian $\mathbf{J}$ is positive semi-definite, that is $\det(\mathbf{J}) \geq 0$. It is strictly convex if $\mathbf{J}$ is positive definite, thus $\det(\mathbf{J}) > 0$.

A.4 Global Solution with Strictly Convex Integrand

Proposition 1. Given the variational problem, minimize

$$H(q) = \int_{\Omega} L(q(\vec{x}), q'(\vec{x}), \vec{x}) \; d\vec{x} \tag{A.10}$$

with $L : \mathbb{R}^m \times \mathbb{R}^m \times \mathbb{R}^n \to \mathbb{R}$ of type C^2 and $q : \Omega \to \mathbb{R}^m$, $\Omega \subset \mathbb{R}^n$, the boundary conditions are free such that $L_{q'}(q(\vec{\xi}), q'(\vec{\xi}), \vec{\xi}) = 0$ with $\vec{\xi} \in \partial\Omega$.

If the integrand $L(q, q', \vec{x})$ is continuous and strictly convex with respect to the functions (q, q'), then a candidate function q_0 that fulfills the boundary conditions and the Euler–Lagrange equations is a global solution of the problem.

Proof. In the following the short notation $L(q, q', \vec{x}) = L(q)$ and $L(q_0, q'_0, \vec{x}) = L(q_0)$, respectively, are used. The idea of the proof is to show that the difference

$$H(q) - H(q_0) = \int_{\Omega} L(q) - L(q_0) \; d\vec{x} \tag{A.11}$$

is always larger than zero, except for $q \equiv q_0$. Since L is strictly convex, the following inequality holds according to Lemma 1:

$$L(q) > L(q_0) + \text{grad}(L(q_0))(q - q_0)$$

$$> L(q_0) + L_q(q_0)(q - q_0) + L_{q'}(q_0)(q' - q_0'). \tag{A.12}$$

Thus, a lower bound for the difference (A.11) can be formulated as

$$H(q) - H(q_0) > \int_{\Omega} L_q(q_0)(q - q_0) + L_{q'}(q_0)(q' - q_0') \, d\vec{x}. \tag{A.13}$$

Partial integration of the second term under the integral

$$\int_{\Omega} L_{q'}(q_0)(q' - q_0') \, d\vec{x} = L_{q'}(q_0)(q - q_0)\Big|_{\partial\Omega} - \int_{\Omega} \frac{d}{d\vec{x}} L_{q'}(q_0)(q - q_0) \, d\vec{x} \tag{A.14}$$

finally leads to

$$H(q) - H(q_0) > \int_{\Omega} \left[L_q(q_0)(q - q_0) - \frac{d}{d\vec{x}} L_{q'}(q_0)(q - q_0) \, d\vec{x} \right] + L_{q'}(q_0)(q - q_0)\Big|_{\partial\Omega}.$$
$$\tag{A.15}$$

The integrand in (A.15) describes the Euler–Lagrange equation which is a necessary condition for a weak solution of the optimization problem. The integral is zero because q_0 is a candidate solution of the problem and thus obeys the Euler–Lagrange equations. The second term is also zero according to the boundary condition. Because

$$H(q) - H(q_0) > 0 \qquad \forall \, q \in \mathbb{R}^m, \tag{A.16}$$

it follows that the energy has a single minimum at q_0.   □

Appendix B

Simulation Methods

The simulation results presented in Chapter 4 are gained by numerical integration of the partial differential equations describing the dynamics of the analog networks. Integration was performed using the explicit Euler method. Step size and stopping criteria were chosen sufficiently small because processing time was not crucial. The network size was determined as equally large as the spatial resolution of the applied image sequence. Although mostly displayed on a sub-sampled grid (see *e.g.* Figure 4.7), simulation results were always performed on the entire image array. Networks were randomly initialized before computing the first frame. If there were subsequent frames to calculate, no re-initialization took place in between frames.

B.1   Spatio-temporal Gradient Estimation

Discretization effects were reduced by presmoothing the image sequences in the spatial and temporal domain [Bertero et al. 1987]. The image sequences were convolved with an isotropic Gaussian kernel of fixed width σ. Depending on the image resolution, the kernel size was chosen differently: $\sigma = 0.25$ pixels for the triangle and the tape-rolls sequence and $\sigma = 0.5$ pixels for the other image sequences. After presmoothing, the gradients in each direction were computed as the symmetric difference between the two nearest neighbors.

B.2   Image Sequences

The triangle sequence

This sequence consists of 20 frames with 64×64 pixel resolution and 8-bit gray-scale encoding. It shows a white triangle moving with a speed of 1 pixel/frame horizontally to the right. The brightness values are normalized to the maximal brightness difference between the darkest points in the background and the surface of the triangle. The contrast of the stationary sinusoidal plaid pattern in the background is half the maximal figure–ground contrast.

Analog VLSI Circuits for the Perception of Visual Motion   A. A. Stocker
© 2006 John Wiley & Sons, Ltd

The tape-rolls sequence

This sequence is a real image sequence of two tape-rolls rolling in opposite directions on an office desk. It consists of 20 frames with a spatial resolution of 64×64 pixels and has 6-bit gray-level depth. The frame rate is 15 frames/s.

The Yosemite sequence

The Yosemite sequence simulates an artificial flight over the Yosemite Valley in California, USA and is based on real topological data overlaid with artificially rendered textures. The clouds in the background are completely fractal. The sequence was generated by Lynn Quann at SRI and has been extensively used as a test sequence for optical flow methods. The sequence consists of 16 frames with a spatial resolution of 316×252 pixels with 8-bit gray-level encoding. Simulations were performed on an image partition containing 256×256 pixels.

The Rubic's cube sequence

The Rubic's cube sequence is another, widely used test sequence for visual motion algorithms. It shows a Rubic's cube on a rotating turn-table. The sequence consists of 21 frames each with a resolution of 256×240 pixels and 8-bit gray-level depth. In simulations, an image partition with a size of 240×240 pixels was used.

The Hamburg taxi sequence

The sequence shows a typical traffic scene at a crossroad, seen out of the window of a high building. There are several motion sources present, including multiple cars and a pedestrian. The original sequence contains 21 frames with 190×256 pixel resolution. In simulations, an image partition of 190×190 pixels was applied.

Where to download?

The "tape-rolls" sequence is available and can be downloaded from the book's website http://wiley.com/go/analog. The other sequences can be found in various archives on the web. Up-to-date links to these archives are also provided on the book's website.

Appendix C

Transistors and Basic Circuits

C.1 Large-signal MOSFET Model

The *metal–oxide–semiconductor field-effect transistor (MOSFET)* is the basic computational device of the implementation of the presented analog networks. Standard *complementary metal–oxide–semiconductor (CMOS)* process technology provides two complementary types of MOSFETs, the native and the well transistor. Labeling of the two types is often based on their channel type: thus the n-channel transistor is called the *nFET* and the p-channel transistor the *pFET*. In a typical p-type substrate process, the nFET is equivalent to the native and the pFET to the well transistor.

Schematic symbols for both transistors can vary in the literature. In this book symbols are used as depicted in Figure C.1. A MOSFET has four terminals called *gate (g)*, *bulk (b)*, *drain (d)*, and *source (s)*. It is important to note, however, that it is a symmetric device because the drain and source are not intrinsically defined by the device. The definition only depends on the voltage distribution on the two terminals. The source is defined as the terminal, where the net current of majority carriers originates. In Figure C.1, I_{ds} indicates the positive channel currents of the transistors in a typical p-type substrate process.

Sub-threshold regime (weak inversion)

In sub-threshold, the channel current I_{ds} is in first order defined by the difference of the diffusion current densities in the *forward* and the *reverse* direction, thus

$$I_{ds} = I_0 \frac{W}{L} \left[\underbrace{\exp\left(-\frac{q}{kT}(\kappa V_g - V_s)\right)}_{\text{forward current}} - \underbrace{\exp\left(-\frac{q}{kT}(\kappa V_g - V_d)\right)}_{\text{reverse current}} \right] \qquad (\text{C.1})$$

where q is the elementary charge of the majority carriers (i.e. for the nFET $q = -1.6022 \cdot 10^{-19}$ C), k the Boltzmann constant, T the temperature, and $\kappa < 1$ a parameter that accounts for the efficacy of the gate voltage V_g on the channel potential. The source V_s, gate V_g, and drain V_d voltages are all referenced to the bulk potential (see Figure C.1). κ can be

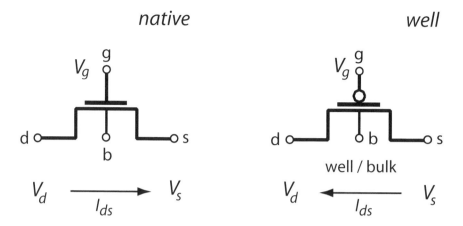

Figure C.1 *Symbolic representation of the native and the well MOSFET.*

approximated by the capacitor ratio $\kappa \approx C_{ox}/(C_{ox} + C_{depletion})$, where C_{ox} is the capacitance of the gate–oxide and $C_{depletion}$ is the capacitance of the depletion region under the channel. $C_{depletion}$ inversely depends on the width of the depletion layer. Thus, κ is not constant but increases with increasing gate–bulk potential.

For a drain–source potential difference $V_{ds} > 4kT/q \approx 100$ mV the reverse current can be neglected and the transistor is said to be in *saturation* where

$$I_{sat} = I_0 \frac{W}{L} \exp\left(-\frac{q}{kT}(\kappa V_g - V_s)\right). \tag{C.2}$$

Above-threshold regime (strong inversion)

In strong inversion, the channel current consists predominantly of drift current. It can be described as a quadratic function of the gate voltage with a reverse and forward component

$$I_{ds} = \frac{\beta}{2\kappa}\big[\underbrace{(\kappa(V_g - V_{T0}) - V_s)^2}_{\text{forward current}} - \underbrace{(\kappa(V_g - V_{T0}) - V_d)^2}_{\text{reverse current}}\big]. \tag{C.3}$$

V_{T0} is the threshold voltage and $\beta = \mu\, C_{ox} W/L$. The parameter κ no longer has a clear physical interpretation as in the sub-threshold regime. It simply represents a slope factor and typically $0.5 < \kappa < 1$. The transistor is said to be in saturation if the reverse current vanishes, thus if the *saturation condition* $V_d > \kappa(V_g - V_{T0})$ holds. In that case Equation (C.3) simplifies to

$$I_{ds} = \frac{\beta}{2\kappa}\left((\kappa(V_g - V_{T0}) - V_s)^2\right). \tag{C.4}$$

Channel-length modulation – the Early effect

The saturated channel current of the MOSFET is linearly dependent on the width-to-length ratio W/L of the transistor gate as expressed by Equations (C.1) and (C.2). Although primarily determined by the physical size of the gate, the effective transistor length, L, also depends on the width of the depletion region at the drain and source junctions. The depletion width increases with increasing reverse bias of the junctions. An effective drain conductances g_{ds} can be defined that accounts for the voltage modulation of the channel length. Thus, the sub-threshold channel current can be rewritten as

$$I_{ds} = I_{sat} + g_{ds} V_{ds},$$ (C.5)

where the drain conductance is defined as

$$g_{ds} = \frac{\partial I_{sat}}{\partial V_{ds}} = \frac{\partial I_{sat}}{\partial L} \frac{\partial L}{\partial V_{ds}}.$$ (C.6)

In a first approximation the channel-length modulation is negatively proportional to the drain potential V_{ds}, hence

$$\frac{\partial L}{\partial V_{ds}} = -c_0$$ (C.7)

where c_0 [m/V] is a constant for a given process and a given transistor length. With

$$\frac{\partial I_{sat}}{\partial L} = -\frac{I_{sat}}{L}$$ (C.8)

the drain conductance becomes

$$g_{ds} = I_{sat} \frac{1}{V_{Early}}, \qquad \text{where} \qquad \frac{1}{V_{Early}} = \frac{c_0}{L}.$$ (C.9)

The Early voltage, V_{Early}, is the hypothetical drain voltage where the extrapolated sub-threshold drain currents all intersect the voltage axis. It is named after Jim Early who first described this effect in bipolar transistors [Early 1952]. Taking into account the effect of channel-length modulation, the total channel current then becomes

$$I_{ds} = I_{sat} \left(1 + \frac{V_{ds}}{V_{Early}}\right),$$ (C.10)

with I_{sat} as given in (C.2).

C.2 Differential Pair Circuit

Consider the differential pair circuit in Figure C.2. In the first instance, all transistors are assumed to be in saturation. For simplicity, any second-order effects (e.g. Early effect) are neglected. All voltages are given in units of kT/q and transistors are assumed to be square ($W/L = 1$).

Sub-threshold regime

The sum of the currents in the two legs of the differential pair equals the bias current provided by M_b, thus

$$I_1 + I_2 = I_b \qquad \text{with} \qquad I_b = I_0 \exp(\kappa V_b).$$ (C.11)

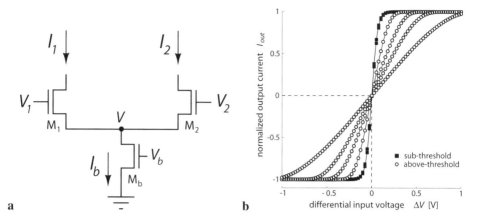

a   **b**

Figure C.2 *Differential pair circuit.*
(a) Circuit diagram. (b) Normalized output currents $I_{out} = (I_1 - I_2)/(I_1 + I_2)$ as functions of the differential input voltage $\Delta V = V_1 - V_2$ for sub- and above-threshold operation. In sub-threshold operation (threshold voltage $V_{th} = 0.79$ V), the *normalized* output currents perfectly overlap because the transconductance linearly scales with the bias current. Thus, the linear range is independent of the bias current (square markers). Above threshold (bias voltage $V_b = [1, 1.25, 1.5, 2$ V]), however, the transconductance only scales with the square root of the bias current and the linear range increases accordingly.

They are given as

$$I_1 = I_0 \exp(\kappa V_1 - V) \qquad \text{and} \qquad I_2 = I_0 \exp(\kappa V_2 - V). \qquad (C.12)$$

Applying (C.12) in (C.11) permits us to solve for V and to reformulate (C.12) as

$$I_1 = I_b \frac{\exp(\kappa V_1)}{\exp(\kappa V_1) + \exp(\kappa V_2)} \qquad \text{and} \qquad I_2 = I_b \frac{\exp(\kappa V_2)}{\exp(\kappa V_1) + \exp(\kappa V_2)}. \qquad (C.13)$$

By multiplying the denominator and numerator of both terms with $\exp(-\kappa(V_1 + V_2)/2)$, the differential output current simplifies to

$$I_{out} = I_1 - I_2 = I_b \tanh\left(\frac{\kappa(V_1 - V_2)}{2}\right). \qquad (C.14)$$

The differential pair is a useful transconductance circuit that provides a well-behaved functional relation between a differential input voltage $\Delta V = V_1 - V_2$ and a bi-directional output current I_{out}. The linear range,[1] however, is limited to $\pm 1.1\,kT/q$ which is roughly ± 30 mV at room temperature. Since I_b appears only as a scaling factor in (C.14) the linear range is independent of the bias current and stays constant. Therefore, the transconductance of the differential pair

$$g_m = \left.\frac{\partial I_{out}}{\partial \Delta V}\right|_{\Delta V = 0} = \frac{I_b \kappa}{2} \qquad \text{with} \qquad \Delta V = V_1 - V_2 \qquad (C.15)$$

---

[1] Here, linear range is defined as the voltage range within which the output current I_{out} does not deviate more than 5% from linearity.

is linear in the bias current. Equation (C.14) is valid as long as the bias transistor is in saturation and I_b is approximately constant. If we assume the extreme case $V_1 \gg V_2$, then all the bias current flows though M_1. To keep M_b in saturation ($V \geq 4kT/q$), the following input condition must hold:

$$\max(V_1, V_2) > V_b + \frac{4kT}{q}. \tag{C.16}$$

Above-threshold regime

When operating the differential pair circuit in strong inversion, the channel currents of transistors M_1, M_2 and M_b are accurately described by Equation (C.4). For reasons of simplicity, β and κ are assumed to be equal and constant for all transistors. Then, the currents in both legs of the differential pair can be described as

$$I_1 = \frac{\beta}{2\kappa}(\kappa(V_1 - V_{T0}) - V)^2 \quad \text{and} \quad I_2 = \frac{\beta}{2\kappa}(\kappa(V_2 - V_{T0}) - V)^2. \tag{C.17}$$

Applying Equations (C.17) to the current equilibrium

$$I_b = I_1 + I_2 \quad \text{with} \quad I_b = \frac{\beta\kappa}{2}(V_b - V_{T0})^2 \tag{C.18}$$

leads to a quadratic expression in V, which has the following solutions:

$$V = \frac{\kappa}{2}\left((V_1 - V_{T0}) + (V_2 - V_{T0}) \pm \sqrt{\frac{4I_b}{\beta\kappa} - \Delta V^2}\right) \tag{C.19}$$

with $\Delta V = V_1 - V_2$. Since V is preferably lower than V_1 or V_2, only the smaller of the two solutions is considered. Substitution of V in Equations (C.17) leads to

$$I_1 = \frac{\beta\kappa}{8}\left(\Delta V + \sqrt{\frac{4I_b}{\beta\kappa} - \Delta V^2}\right)^2 \quad \text{and} \quad I_2 = \frac{\beta\kappa}{8}\left(-\Delta V + \sqrt{\frac{4I_b}{\beta\kappa} - \Delta V^2}\right)^2. \tag{C.20}$$

Finally, the output current is

$$I_{out} = \frac{\beta\kappa}{2}\Delta V\sqrt{\frac{4I_b}{\beta\kappa} - \Delta V^2}. \tag{C.21}$$

Under which conditions is Equation (C.21) a good approximation of the above-threshold behavior of the differential pair? First, M_b must be in saturation. According to (C.4) this is the case if $V > \kappa(V_b - V_{T0})$. Since V is lowest if the complete bias current flows through either one of the two legs, the lower *saturation limit* for one of the input voltages follows as

$$\max(V_1, V_2) > V_b + (V_b - V_{T0}). \tag{C.22}$$

Second, transistors M_1 and M_2 must remain in above-threshold operation. This condition determines the upper limit for the absolute value of the differential input voltage $|\Delta V|$. For

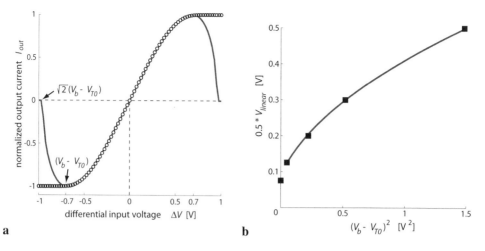

a   **b**

Figure C.3 *Differential pair: above-threshold characteristics.*
(a) The measured normalized output current I_{out} is shown as a function of the differential
input voltage $\Delta V = V_1 - V_2$ for $V_b = 1.5$ V. The bold curve is a least-squares error fit
according to Equation (C.21). The fit is accurate within input limits $\Delta V = \pm(V_b - V_{T0})$
(with $V_{T0} = 0.79$ V). When the square root in Equation (C.21) turns negative, the real part
of the fit is zero. This occurs at $\Delta V = \pm 0.98$ V which is approximately $\sqrt{2}(V_b - V_{T0})$.
(b) In above-threshold operation, the linear range (defined as 5% deviation limit) increases
inversely proportionally to the normalized transconductance g_m/I_b, thus proportionally to
the square root of the bias current. The fit (bold line) matches the data well.

example, assume that M_2 is operating right at threshold,[2] which is the case if $\kappa(V_2 - V_{T0}) -$
$V = 0$. Furthermore, any residual sub-threshold current in M_2 is neglected. Substitution of
$V = \kappa(V_2 - V_{T0})$ in I_1 (C.17) with the approximation $I_1 = I_b$ leads to the second condition,
defining the *input limit*

$$\Delta V \leq (V_b - V_{T0}). \tag{C.23}$$

Equation (C.21) is not a good approximation over the entire input range for which its
square root is positive. As given by (C.23), the valid input range is smaller by a factor of
$\sqrt{2}$. Figure C.3 shows the measured output current and a least-squares error fit based on
Equation (C.21). The input limit is not important in practice because the real circuit, unlike
the function (C.21), is very well-behaved: beyond the input limit (C.23) the output current
saturates and the total bias current I_b is flowing through either transistor M_1 or M_2.

In above-threshold operation, the transconductance of the differential pair is given as

$$g_m = \sqrt{I_b \beta \kappa}. \tag{C.24}$$

Clearly, the transconductance is no longer proportional to the bias current I_b, so the linear
range increases as I_b gets larger. Under the assumption that the characteristic shape of the

---

[2]Correctly speaking, at threshold the transistors are operating in *moderate inversion*. This region roughly covers
are range of 200 mV around the threshold voltage. Thus, the above assumption slightly overestimates the maximal
input range that ensures above-threshold operation.

normalized output currents remains similar for different bias current levels, the limits for linearity occur at constant current levels. Hence, the linear range is inversely proportional to the normalized transconductance, thus

$$\Delta V_{\text{linear}} \propto \left(\frac{g_m}{I_b}\right)^{-1} \propto \sqrt{I_b}. \tag{C.25}$$

Figure C.3b shows the measured linear range and a least-squares error fit according to (C.25).

C.3 Transconductance Amplifiers

Signal amplification is an important step in signal processing systems. The two most basic transconductance amplifier circuits are briefly discussed here. The term *transconductance* refers to the fact that the input and output terminals of these amplifiers are electronically isolated from each other; there is no direct current flowing between these nodes. In the following analysis, all voltages are in units of kT/q and transistors are assumed to be square ($W/L = 1$).

C.3.1 Inverting amplifier

Connecting two transistors of opposite type at their drains and holding their sources at fixed potentials provides a simple inverting transconductance amplifier. This amplifier is used, for example, in the adaptive photoreceptor circuit (Figure 5.2a, page 98) and the clamped-capacitor differentiator circuit (Figure 5.3, page 101) discussed previously.

The simplest configuration is depicted in Figure C.4b. The input voltage V_{in} is applied to the gate of the nFET while the pFET is biased by a fixed gate voltage V_b. Both transistors must source the same current. Thus, a small change in V_{in} results in a large opposite change in V_{out}, the only terminal that is not controlled. For small signal changes, the amplifier gain can be computed from the Early effect (C.10) as

$$A \equiv \frac{\partial V_{out}}{\partial V_{in}} = -\kappa_N \frac{V_{E,N} V_{E,P}}{V_{E,P} + V_{E,N}}. \tag{C.26}$$

An equivalent result can be obtained for the inverse circuit configuration, where the input V_{in} is applied to the gate of the pFET and the nFET is biased by a fixed voltage V_b.

Now consider the case where the input is applied to both gates (Figure C.4c). The gain of this amplifier is given as

$$A \equiv \frac{\partial V_{out}}{\partial V_{in}} = -(\kappa_N + \kappa_P) \frac{V_{E,N} V_{E,P}}{V_{E,P} + V_{E,N}} \tag{C.27}$$

and is even higher than in the previous configuration. This circuit is often referred to simply as an *inverter* and is one of the fundamental logic building block in digital circuits.

C.3.2 Differential amplifier

The symbol and the circuit schematics of the simple differential transconductance amplifier are shown in Figure C.5. The circuit consists of a differential pair combined with

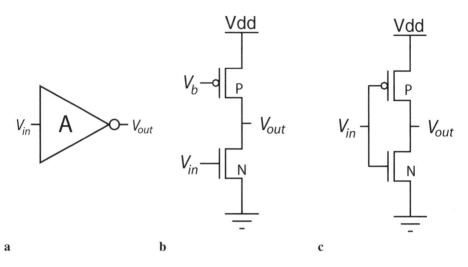

a   **b**   **c**

Figure C.4 *Inverting transconductance amplifier.*
(a) Symbol. (b) Schematics of the inverting amplifier with bias transistor. (c) Full inverter circuit.

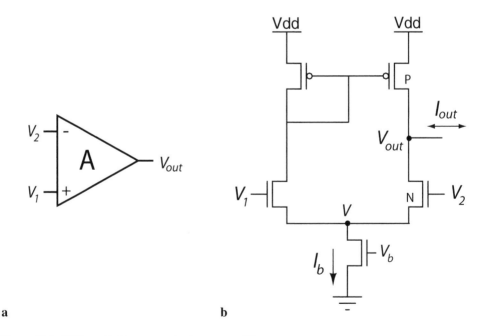

a   **b**

Figure C.5 *Differential transconductance amplifier.*
(a) Symbol. (b) Schematics of the differential transconductance amplifier in its minimal configuration.

a current mirror. Thus, in a first approximation (ideal current mirror) the output current of the amplifier and its transconductance are as derived in the previous analysis of the differential pair.

The gain of the differential amplifier circuit is defined as

$$A \equiv \frac{\partial V_{out}}{\partial (V_1 - V_2)} = \frac{g_m}{g_o} \tag{C.28}$$

where g_m is its transconductance and g_o the output conductance at node V_{out}. The output conductance is mainly determined by the Early voltages of the transistors N and P. For different operational regimes the gain of the amplifier is different. In sub-threshold, with the transconductance given as (C.15) and the output conductance according to (C.6), the gain can be computed as

$$A = \kappa (V_{E,N} + V_{E,P}) \tag{C.29}$$

where κ is the slope factor of the nFETs of the differential pair.

The transconductance changes above threshold, and thus so does the gain. With (C.24) and (C.28) the above-threshold gain is given as

$$A = 2\sqrt{\frac{\beta \kappa}{I_b}} (V_{E,N} + V_{E,P}) . \tag{C.30}$$

Note that the output conductance changes also because the Early voltages of the transistors are different when operating in the above-threshold regime.

C.4   References

This appendix provides only a limited and selected discussion on CMOS transistors and circuits. Excellent textbooks and papers exist, however, that cover in great detail many aspects of analog integrated circuit design. In the following, I would like to point out a few selected textbooks.

Sze's book *Physics of Semiconductor Devices* [Sze 1981] covers in detail the physics of CMOS transistors and circuits. It explains pn junctions as well as simple devices for phototransduction. This is a very good reference text.

Another good source is Gray and Meyer's textbook *Analysis and Design of Analog Integrated Circuits* [Gray and Meyer 1993]. It focuses less on semiconductor physics, but discusses several transistor models. It also specifically addresses bipolar transistors. The analysis of simple amplifier circuits and discussions on noise complete this book.

Carver Mead's *Analog VLSI and Neural Systems* [Mead 1989] is a classic and describes many of the circuits that serve as the basic building blocks in neuromorphic analog VLSI systems. The book is short on explaining the underlying device physics but covers many static and dynamic circuits. It also includes descriptions of complete neuromorphic systems such as the silicon retina. This book is an enjoyable read and introduces the reader to the ideas and motivations for building neurally inspired analog circuits. Liu and colleagues' *Analog VLSI: Circuits and Principles* [Liu et al. 2002] represents in many aspects an updated version of Mead's book. It is more detailed in describing transistor physics and also includes many of the known transistor models. It introduces a variety of static and dynamic analog circuits, also including complete chapters on floating-gate and photoreceptor circuits.

Appendix D

Process Parameters and Chips Specifications

D.1 AMS 0.8 μm BiCMOS Process

The *AustriaMicroSystems AMS* 0.8 μm BiCMOS process is a mixed analog/digital process and offers the implementation of a generic npn–bipolar transistor. The process provides two metal layers and two layers of low ohmic polysilicon (poly) to form poly/poly capacitors. The substrate is acceptor doped (p−) and the gates consist of heavily donor-doped polysilicon (n+). This process was available through the multi-project-wafer (MPW) service provided by Europractice (www.europractice.com). The most important process parameters are listed in Table D.1.

Table D.1 AMS 0.8 μm BiCMOS process parameters.

MOSFET transistor	NMOS	PMOS	units
Effective substrate doping	74×10^{15}	28×10^{15}	cm^{-3}
Minimum drawn gate length	0.8	0.8	μm
Oxide thickness	16	16	nm
Threshold voltage 20/20	0.80	−0.84	V
Transconductance	100	35	μ A/V^2
Body factor 20/20	0.74	0.45	$\sqrt{V}$
Bipolar transistor	*NPN*		
Beta	100		–
Early-voltage V_{Early}	32		V
Transit frequency F_t	12		GHz
Thin oxide poly/poly capacitor	1.8		fF/μm^2
Operating voltage	2.5–5.5		V

Analog VLSI Circuits for the Perception of Visual Motion A. A. Stocker
© 2006 John Wiley & Sons, Ltd

D.2   Chip Specifications

Table D.2   Specification of the three visual motion chips presented in this book.

The smooth optical flow chip	
Die size(s)	$2.55 \times 2.55 \, \text{mm}^2 \, / \, 4.72 \times 4.5 \, \text{mm}^2$
Array size(s) (motion units)	$10 \times 10 / 30 \times 30$
Pixel size	$124 \times 124 \, \mu\text{m}^2$
Active elements	119
Power dissipation	$52 \, \mu\text{W}$
Outputs (scanned)	optical flow, photoreceptor
Typical read-out speed	1000 frames/s
The motion segmentation chip	
Die size	$2.7 \times 2.95 \, \text{mm}^2$
Array size (motion units)	11×11
Pixel size	$170 \times 170 \, \mu\text{m}^2$
Active elements (per pixel)	203
Fill factor[a]	2.1%
Power dissipation (per pixel)	$80 \, \mu\text{W}$
Outputs signals (scanned)	5: photoreceptor, optical flow, P and Q
Max. read-out speed	$\approx 10,000$ frames/s
Typical read-out speed	1000 frames/s
The motion selection chip	
Die size	$2.7 \times 2.9 \, \text{mm}^2$
Array size (motion units)	15×15
Pixel size	$132 \times 132 \, \mu\text{m}^2$
Fill factor	4.1%
Read/write cycle	typically $< 5 \, \text{ms}$

[a]Previously specified as 4% which is not quite correct [Stocker 2002]. This larger value represents the total area over which visual information is averaged.

References

Adelson, E. and J. Bergen (1985, February). Spatiotemporal energy models for the perception of motion. *Journal of the Optical Society of America 2*(2), 284–299.

Adelson, E. and J. Movshon (1982, December). Phenomenal coherence of moving visual patterns. *Nature 300*(9), 523–525.

Allman, J., F. Miezin, and E. McGuinness (1985). Stimulus specific responses from beyond the classical receptive field: Neurophysiological mechanisms for local-global comparisons in visual neurons. *Annual Reviews in Neuroscience 8*, 407–430.

Anandan, P. (1989). A computational framework and an algorithm for the measurement of visual motion. *International Journal of Computer Vision 2*, 283–310.

Anderson, C., H. Nover, and G. DeAngelis (2003). Modeling the velocity tuning of macaque MT neurons. *Journal of Vision/VSS abstract 3*(9), 404a.

Andreou, A. and K. Strohbehn (1990). Analog VLSI implementation of the Hassenstein-Reichardt-Poggio models for vision computation. *Proceedings of the 1990 International Conference on Systems, Man and Cybernetics*, Los Angeles.

Andreou, A., K. Strohbehn, and R. Jenkins (1991). Silicon retina for motion computation. In *Proceedings of the IEEE International Symposium on Circuits and Systems*, Singapore, pp. 1373–1376.

Arreguit, X., A. van Schaik, F. Bauduin, M. Bidiville, and E. Raeber (1996, December). A CMOS motion detector system for pointing devices. *IEEE Journal of Solid-State Circuits 31*(12), 1916–1921.

Aubert, G., R. Deriche, and P. Kornprobst (1999). Computing optical flow via variational techniques. *SIAM Journal of Applied Mathematics 60*(1), 156–182.

Barlow, H. and W. Levick (1965). The mechanism of directionally selective units in rabbit's retina. *Journal of Physiology 178*, 477–504.

Barron, J., D. Fleet, and S. Beauchemin (1994). Performance of optical flow techniques. *International Journal of Computer Vision 12*(1), 43–77.

Becanovic, V., S. Kubina, and A. Stocker (2004, December). An embedded neuromorphic vision system suitable for mobile robotics applications. In *11th IEEE International Conference on Mechatronics and Machine Vision in Practice*, Macau, pp. 127–131.

Benson, R. and T. Delbrück (1992). Direction-selective silicon retina that uses null inhibition. In D. Touretzky (Ed.), *Neural Information Processing Systems 4*, pp. 756–763. Cambridge, MA: MIT Press.

Bertero, M., T. Poggio, and V. Torre (1987, May). Ill-posed problems in early vision. Technical Report 924, MIT AI Lab, Cambridge, MA.

Bird, C., G. Henning, and F. Wichmann (2002, July). Contrast discrimination with sinusoidal gratings of different spatial frequency. *Journal of the Optical Society of America A 19*(7), 1267–1273.

Braddick, O. (1993). Segmentation versus integration in visual motion processing. *Trends in Neuroscience 16*(7), 263–268.

Bradley, D. and R. Andersen (1998, September). Center-surround antagonism based on disparity in primate area MT. *Journal of Neuroscience 15*, 7552–7565.

Bremmer, F. and M. Lappe (1999, July). The use of optical velocities for distance discrimination and reproduction during visually simulated self-motion. *Experimental Brain Research 127*(1), 33–42.

Britten, K., W. Newsome, M. Shadlen, S. Celebrini, and A. Movshon (1996). A relationship between behavioral choice and the visual responses for neurons in macaque MT. *Visual Neuroscience 13*, 87–100.

Bronstein, I. and K. Smendjajew (1996). *Teubner-Taschenbuch der Mathematik*. Stuttgart, Leibzig: B.G. Teubner.

Bülthoff, H., J. Little, and T. Poggio (1989, February). A parallel algorithm for real-time computation of optical flow. *Nature 337*(9), 549–553.

Burke, D. and P. Wenderoth (1993). The effect of interactions between one-dimensional component gratings on two-dimensional motion perception. *Vision Research 33*(3), 343–350.

Camus, T. (1994, September). *Real-Time Optical Flow*. Ph.D. thesis, Brown University, Dept. of Computer Science.

Cesmeli, E. and D. Wang (2000, July). Motion segmentation based on motion/brightness integration and oscillatory correlation. *IEEE Transactions on Neural Networks 11*(4), 935–947.

Chahine, M. and J. Konrad (1995, November). Estimation and compensation of accelerated motion for temporal sequence interpolation. *Signal Processing, Image Communication 7*, 503–527.

Chance, F., S. Nelson, and L. Abbott (1999). Complex cells as cortically amplified simple cells. *Nature Neuroscience 2*, 277–282.

Chklovskii, D. (2004). Exact solution for the optimal neuronal layout problem. *Neural Computation 16*, 2067–2078.

Cichocki, A. and R. Unbehauen (1993). *Neural Networks for Optimization and Signal Processing* (3rd ed.). Chichester: John Wiley & Sons, Ltd.

Cragg, B. and H. Temperley (1955). Memory: The analogy with ferromagnetic hysteresis. *Brain 78*, 304–316.

Delbrück, T. (1991). "Bump" circuits for computing similarity and dissimilarity of analog voltages. In *Proceedings of the International Joint Conference on Neural Networks*, Seattle, Volume 1, pp. 475–479.

Delbrück, T. (1993a, May). Bump circuits. Technical Report CNS Memo 26, Caltech, Pasadena, CA.

Delbrück, T. (1993b). *Investigations of analog VLSI visual transduction and motion processing.* Ph.D. thesis, Dept. of Computational and Neural Systems, California Institute of Technology.

Delbrück, T. (1993c, May). Silicon retina with correlation-based velocity-tuned pixels. *IEEE Transactions on Neural Networks 4*(3), 529–541.

Delbrück, T. and C. Mead (1989). An electronic photoreceptor sensitive to small changes in intensity. In D. Touretzky (Ed.), *Advances in Neural Information Processing Systems*, Volume 1, pp. 720–726. San Francisco: Morgan Kaufmann.

Delbrück, T. and C. Mead (1994). Analog VLSI phototransduction by continuous-time, adaptive, logarithmic photoreceptor circuits. Technical Report 30, Caltech Computation and Neural Systems Program.

Deutschmann, R. and C. Koch (1998a). An analog VLSI velocity sensor using the gradient method. In *Proceedings of the IEEE International Symposium on Circuits and Systems*, Monterey, CA, pp. 649–652.

Deutschmann, R. and C. Koch (1998b). Compact real-time 2-D gradient-based analog VLSI motion sensor. In *Proceedings of the SPIE International Conference on Advanced Focal Plane Arrays and Electronic Cameras*, Volume 3410, pp. 98–108.

Duchon, J. (1977). Splines minimizing rotation-invariant semi-norms in Sobolev spaces. In A. Dold and B. Eckmann (Eds.), *Constructive Theory of Functions of Several Variables*, pp. 85–100. Springer Verlag: Berlin.

Duffy, C. (2000). Optic flow analysis for self-movement perception. In M. Lappe (Ed.), *Neural Processing of Optic Flow*, pp. 199–218, San Diego, CA: Academic Press.

Duffy, C. and R. Wurtz (1991a). Sensitivity of MST neurons to optic flow stimuli. 1. A continuum of response selectivity to large-field stimuli. *Journal of Neurophysiology 65*(6), 1329–1345.

Duffy, C. and R. Wurtz (1991b). Sensitivity of MST neurons to optic flow stimuli. 2. Mechanisms of response selectivity revealed by small-field stimuli. *Journal of Neurophysiology 65*(6), 1346–1359.

Duffy, C. and R. Wurtz (1995). Response of monkey MST neurons to optic flow stimuli with shifted centers of motion. *Journal of Neuroscience 15*, 5192–5208.

Early, J. (1952). Effects of space-charge layer widening in junction transistors. *Proceedings of the Institute of Radio Engineers 40*, 1401–1406.

Etienne-Cummings, R. (1994). *Biologically Motivated Analog VLSI Systems for Optomotor Tasks.* Ph.D. thesis, University of Pennsylvania.

Etienne-Cummings, R., S. Fernando, N. Takahashi, V. Shotonov, J. Van der Spiegel, and P. Müller (1993, December). A new temporal domain optical flow measurement technique for focal plane VLSI implementation. In *Computer Architectures for Machine Perception*, New Orleans, LA, pp. 241–250.

Etienne-Cummings, R., J. Van der Spiegel, and P. Mueller (1997, January). A focal plane visual motion measurement sensor. IEEE *Transactions on Circuits and Systems I 44*(1), 55–66.

Etienne-Cummings, R., J. Van der Spiegel, and P. Mueller (1999, September). Hardware implementation of a visual-motion pixel using oriented spatiotemporal neural filters. *IEEE Transactions on Circuits and Systems 2 46*(9), 1121–1136.

Etienne-Cummings, R., J. Van der Spiegel, N. Takahashi, and P. Mueller (1996, November). VLSI implementation of cortical visual motion detection using an analog neural computer. In *Advances in Neural Information Processing Systems 8*, Denver, CO.

Farid, H. and E. Simoncelli (2003, April). Differentiation of discrete multidimensional signals. *IEEE Transactions on Image Processing 13*(4), 496ff.

Fennema, C. and W. Thompson (1979). Velocity determination in scenes containing several moving objects. *Computer Graphics and Image Processing 9*, 301–315.

Fleet, D. and A. Jepson (1990). Computation of component image velocity from local phase information. *International Journal of Computer Vision 5*(1), 77–104.

Fletcher, R. (1980). *Practical Methods of Optimization*, Volume 1, Chichester John Wiley & Sons Ltd.

Fletcher, R. (1981). *Practical Methods of Optimization*, Volume 2, Chichester John Wiley & Sons Ltd.

Foley, J. and G. Legge (1981). Contrast detection and near-threshold discrimination in human vision. *Vision Research 21*, 1041–1053.

Fukushima, K., Y. Yamaguchi, M. Yasuda, and S. Nagata (1970, December). An electronic model of the retina. *Proceedings of the IEEE 58*(12), 1950–1951.

Gee, A. and R. Prager (1995, January). Limitations of neural networks for solving Traveling Salesman problems. *IEEE Transactions on Neural Networks 6*(1), 280–282.

Gibson, J. (1986). *The Ecological Approach to Visual Perception*. Hillsdale, NJ: Lawrence Erlbaum associates.

Gibson, J. J. (1950). *The Perception of the Visual World*. Boston, MA: Houghton Mifflin.

Gilbert, B. (1968, December). A precise four-quadrant multiplier with subnanosecond response. *IEEE Journal of Solid-State Circuits 3*(4), 363–373.

Gray, P. and R. Meyer (1993). *Analysis and Design of Analog Integrated Circuits* (3rd ed.). New York: John Wiley & Sons, Inc.

Grossberg, S. (1978). Competition, decision and consensus. *Journal of Mathematical Analysis and Applications 66*, 470–493.

Grzywacz, N. and A. Yuille (1990). A model for the estimate of local image velocity by cells in the visual cortex. *Proceedings of the Royal Society of London B*(239), 129–161.

Grzywacz, N. and A. Yuille (1991). Theories for the visual perception of local velocity and coherent motion. In M. S. Landy and J. A. Movshon (Eds.), *Computational Models of Visual Processing*, pp. 231–252. Cambridge, MA: MIT Press.

Hahnloser, R., R. Douglas, M. Mahowald, and K. Hepp (1999, August). Feedback interactions between neuronal pointers and maps for attentional processing. *Nature Neuroscience 2*(8), 746–752.

Harris, J. (1991). *Analog Models for Early Vision*. Ph.D. thesis, California Institute of Technology.

Harris, J. and C. Koch (1989, April). Resistive fuses: Circuit implementations of line discontinuities in vision. *Neural Networks for Computing Workshop*, Snowbird, UT.

Harris, J., C. Koch, and J. Luo (1990a, June). A two-dimensional analog VLSI circuit for detecting discontinuities in early vision. *Science 248*, 1209–1211.

Harris, J., C. Koch, E. Staats, and J. Luo (1990b). Analog hardware for detecting discontinuities in early vision. *International Journal of Computer Vision 4*, 211–223.

Harrison, R. and C. Koch (1998). An analog VLSI model of the fly elementary motion detector. In M. Kearns and S. Solla (Eds.), *Advances in Neural Information Processing Systems 10*, pp. 880–886. Cambridge, MA: MIT Press.

Hassenstein, B. and W. Reichardt (1956). Systemtheoretische Analyse der Zeitreihenfolgen und Vorzeichenauswertung bei der Bewegungsperzeption des Rüsselkäfers Chlorophanus. *Zeitschrift für Naturforschung 11b*, 513–524.

Hebb, D. (1949). *The Organization of Behaviour*. New York: John Wiley & Sons, Inc.

Heeger, D. (1987a). A model for the extraction of image flow. *Journal of the Optical Society of America 4*, 151–154.

Heeger, D. (1987b, August). Model for the extraction of image flow. *Journal of the Optical Society of America 4*(8), 1455–1471.

Heeger, D. J., E. Simoncelli, and J. Movshon (1996, January). Computational models of cortical visual processing. *Proceedings of the National Academy of Sciences USA 93*, 623–627.

Heinzle, J. and A. Stocker (2003). Classifying patterns of visual motion - A neuromorphic approach. In S. T. S. Becker and K. Obermayer (Eds.), *Advances in Neural Information Processing Systems 15*, pp. 1123–1130. Cambridge, MA: MIT Press.

Hertz, J., A. Krogh, and R. Palmer (1991). *Introduction to the theory of neural computation*. Lecture Notes–Santa Fe Institute. Reading, MA: Perseus Books.

Higgins, C. and C. Koch (1997, February). Analog CMOS velocity sensors. *Proceedings of Electronic Imaging, SPIE 3019*.

Higgins, C., R. Deutschmann, and C. Koch (1999). Puhe - based 2-D motion sensor. *IEEE Transaction on Circuits and System 2 46*(6), 677–687.

Hildreth, E. C. (1983). *The Measurement of Visual Motion*. Cambridge, MA: MIT Press.

Hopfield, J. (1982, April). Neural networks and physical systems with emergent collective computational abilities. *Proceedings of the National Academy of Sciences USA 79*, 2554–2558.

Hopfield, J. (1984, May). Neurons with graded response have collective computational properties like those of two-state neurons. *Proceedings of the National Academy of Sciences USA 81*, 3088–3092.

Hopfield, J. and D. Tank (1985). Neural computation of decisions in optimization problems. *Biological Cybernetics 52*, 141–152.

Horiuchi, T., B. Bishofberger, and C. Koch (1994). An analog VLSI saccadic eye movement system. In J. Cowan, G. Tesauro, and J. Alspector (Eds.), *Advances in Neural Information Processing Systems 6*, pp. 582–589. San Francisco: Morgan Kaufmann.

Horiuchi, T., J. P. Lazzaro, A. Moore, and C. Koch (1991). A delay-line based motion detection chip. In R. Lippman, J. Moody, and D. Touretzky (Eds.), *Advances in Neural Information Processing Systems 3*, Volume 3, pp. 406–412 Morgan Kaufmann, Denver, CO.

Horn, B. and B. Schunck (1981). Determining optical flow. *Artificial Intelligence 17*, 185–203.

Horn, B. K. (1986). *Robot Vision*. New York: McGraw-Hill.

Horn, B. K. (1988, December). Parallel networks for machine vision. Technical Report 1071, MIT AI Lab, Cambridge, MA.

Hoyer, P. and A. Hyvarinen (2002, December). Interpreting neural response variability as Monte-Carlo sampling of the posterior. In *Advances in Neural Information Processing Systems NIPS 15*, pp. 277–284. Cambridge, MA: MIT Press.

Huang, L. and Y. Aloimonos (1991, October). Relative depth from motion using normal flow: An active and purposive solution. In *IEEE Workshop on Visual Motion*, pp. 196–203. Los Alamitos, CA: IEEE Computer Society Press.

Hubel, D. and A. Wiesel (1962). Receptive fields, binocular interaction and functional architecture in the cats visual cortex. *Journal of Physiology 160*, 106–154.

Hupe, J., A. James, B. Payne, S. Lomber, P. Girard, and J. Bullier (1998, August). Cortical feedback improves discrimination between figure and background by V1, V2 and V3 neurons. *Nature 394*, 784–787.

Hürlimann, F., D. Kiper, and M. Carandini (2002). Testing the Bayesian model of perceived speed. *Vision Research 42*, 2253–2257.

Hutchinson, J., C. Koch, J. Luo, and C. Mead (1988, March). Computing motion using analog and binary resistive networks. *Computer 21*, 52–64.

Indiveri, G., J. Kramer, and C. Koch (1996). Parallel analog VLSI architectures for computation of heading direction and time-to-contact. In *Advances in Neural Information Processing Systems 8*, pp. 720–726.

Jiang, H.-C. and C.-Y. Wu (1999, February). A 2-D velocity- and direction-selective sensor with BJT-based silicon retina and temporal zero-crossing detector. *IEEE Journal of Solid-State Circuits 34*(2), 241–247.

Kalman, Rudolph, E. (1960). A new approach to linear filtering and prediction problems. *Transactions of the ASME–Journal of Basic Engineering 82*(Series D), 35–45.

Kaski, S. and T. Kohonen (1994). Winner-take-all networks for physiological models of competitive learning. *Neural Networks 7*(6/7), 973–984.

Kastner, S. and L. Ungerleider (2000). Mechanisms of visual attention in the human cortex. *Annual Review of Neurosciences 23*, 315–341.

Kitano, H., M. Asada, Y. Kuniyoshi, I. Noda, and E. Osawa (1997). RoboCup: The robot world cup initiative. In W. L. Johnson and B. Hayes-Roth (Eds.), *Proceedings of the First International Conference on Autonomous Agents (Agents'97)*, New York, pp. 340–347. New York: ACM Press.

Koch, C., J. Marroquin, and A. Yuille (1986). Analog neuronal networks in early vision. *Proceedings of the National Academy of Sciences USA 83*, 4263–4267.

Koch, C., H. Wang, R. Battiti, B. Mathur, and C. Ziomkowski (1991, October). An adaptive multi-scale approach for estimating optical flow: Computational theory and physiological implementation. *IEEE Workshop on Visual Motion*, Princeton, New Jersey, pp. 111–122.

Koch, C., H. Wang, and B. Mathur (1989). Computing motion in the primate's visual system. *Journal of Experimental Biology 146*, 115–139.

Koenderink, J. (1986). Optic flow. *Vision Research 26*(1), 161–180.

Konrad, J. and E. Dubois (1992, September). Bayesian estimation of motion vector fields. *IEEE Transactions on Pattern Analysis and Machine Intelligence 14*(9), 910–927.

Koulakov, A. and D. Chklovskii (2001, February). Orientation preference patterns in mammalian visual cortex: A wire length minimization approach. *Neuron 29*, 519–527.

Kramer, J. (1996, April). Compact integrated motion sensor with three-pixel interaction. *IEEE Transactions on Pattern Analysis and Machine Intelligence 18*(4), 455–460.

Kramer, J., R. Sarpeshkar, and C. Koch (1995). An analog VLSI velocity sensor. In *IEEE International Symposium on Circuits and Systems*, Seattle, WA, pp. 413–416.

Kramer, J., R. Sarpeshkar, and C. Koch (1996). Analog VLSI motion discontinuity detectors for image segmentation. In *IEEE International Symposium on Circuits and Systems*, Atlanta, GA, pp. 620–623.

Kramer, J., R. Sarpeshkar, and C. Koch (1997, February). Pulse-based analog VLSI velocity sensors. *IEEE Transactions on Circuits and Systems 2 44*(2), 86–101.

Krapp, H. (2000). Neuronal matched filters for optic flow processing in flying insects, *Neuronal Processing of Optical Flow*, pp. 93–120. In M. Lappe (Ed.), San Diego, CA: Academic Press.

Lamme, V., B. van Dijk, and H. Spekreijse (1993, June). Contour from motion processing occurs in primary visual cortex. *Nature 363*, 541–543.

Lappe, M. (2000). Computational mechanisms for optic flow analysis in primate cortex. In M. Lappe (Ed.), *Neural Processing of Optic Flow*, pp. 235–268. San Diego, CA: Academic Press.

Lappe, M., F. Bremmer, M. Pekel, A. Thiele, and K.-P. Hoffmann (1996, October). Optic flow processing in monkey STS: A theoretical and experimental approach. *Journal of Neuroscience 16*, (19), 6265–6285.

Lei, M.-H. and T.-D. Chiueh (2002, May). An analog motion field detection chip for image segmentation. *IEEE Transactions on Circuits and Systems for Video Technology 12*(5), 299–308.

Liang, X.-B. and J. Wang (2000, November). A recurrent neural network for nonlinear optimization with a continuously differentiable objective function and bound constraints. *IEEE Transactions on Neural Networks 11*(6), 1251–1262.

Little, J. and A. Verri (1989, March). Analysis of differential and matching methods for optical flow. In *Workshop on Visual Motion*, pp. 173–180. Los Alamitos, CA: IEEE Computer Society Press.

Little, W. and G. Shaw (1975). A statistical theory of short and long term memory. *Behavioral Biology 14*, 115–133.

Liu, S.-C. (1996, June). Silicon model of motion adaptation in the fly visual system. *Proceedings of the 3rd Joint Symposium on Neural Computation*, Paradera, CA.

Liu, S.-C. (1998, November). Silicon retina with adaptive filtering properties. In *Advances in Neural Information Processing Systems 10*, pp. 712–718. Cambridge, MA: MIT Press.

Liu, S.-C. (2000). A neuromorphic aVLSI model of global motion processing in the fly. *IEEE Transactions on Circuits and Systems 2 47*(12), 1458–1467.

Liu, S.-C. and J. Harris (1992). Dynamic wires: An analog VLSI model for object-based processing. *International Journal of Computer Vision 8*(3), 231–239.

Liu, S.-C., J. Kramer, G. Indiveri, T. Delbrück, and R. Douglas (2002). *Analog VLSI: Circuits and Principles*. Cambridge, MA: MIT Press.

Lucas, B. and T. Kanade (1981). An iterative image registration technique with an application to stereo vision. In *Proceedings of the Image Understanding Workshop*, Vancouver, pp. 121–130. DARPA.

Maass, W. (2000). On the computational power of winner-take-all. *Neural Computation 12*, 2519–2535.

Mahowald, M. (1991, April). Silicon retina with adaptive photoreceptor. In *Proceedings of the SPIE Symposium on Electronic Science and Technology: from Neurons to Chips*, Volume 1473, pp. 52–58.

Mahowald, M. and C. Mead (1991, May). The silicon retina. *Scientific American*, 76–82.

Marr, D. (1982). *Vision*. San Francisco: W.H. Freeman.

Marr, D. and T. Poggio (1976, October). Cooperative computation of stereo disparity. *Science 194*, 283–287.

McCulloch, S. and W. Pitts (1943). A logical calculus of the ideas immanent in nervous activity. *Bulletin of Mathematical Biophysics 5*, 115–133.

Mead, A. and T. Delbrück (1991, July). Scanners for visualizing activity of analog VLSI circuitry. Technical Report CNS Memo 11, Caltech, Paradera, CA.

Mead, C. (1989). *Analog VLSI and Neural Systems*. Reading, MA: Addison-Wesley.

Mead, C. (1990, October). Neuromorphic electronic systems. *Proceedings of the IEEE 78*(10), 1629–1636.

Mehta, S. and R. Etienne-Cummings (2004, May). Normal optical flow measurement on a CMOS APS imager. In *IEEE International Symposium on Circuits and Systems*, Vancouver, Volume 4, pp. 848–851.

Memin, E. and P. Perez (1998, May). Dense estimation and object-based segmentation of the optical flow with robust techniques. *IEEE Transactions on Image Processing 7*(5), 703–719.

Mikami, A., W. Newsome, and R. Wurtz (1986, June). Motion selectivity in macaque visual cortex. 1. Mechanisms of direction and speed selectivity in extrastriate area MT. *Journal of Neurophysiology 55*(6), 1308–1327.

Miller, K. and G. Borrows (1999, May). Feature tracking linear optic flow sensor chip. In *IEEE International Symposium on Circuits and Systems*, Orlando, FL, Volume 5, pp. 116–119.

Moini, A., A. Bouzerdoum, K. Eshraghian, A. Yakovleff, X. Nguyen, A. Blanksby, R. Beare, D. Abbott, and R. Bogner (1997, February). An insect vision-based motion detection chip. *IEEE Journal of Solid-State Circuits 32*(2), 279–283.

Moore, A. and C. Koch (1991). A multiplication based analog motion detection chip. In B. Mathur and C. Koch (Eds.), *SPIE Visual Information Processing: From Neurons to Chips*, Volume 1473, pp. 66–75.

Moore, G. (1965, April). Cramming more components onto integrated circuits. *Electronics 38*(8).

Movshon, J., E. Adelson, M. Gizzi, and W. Newsome (1985). The analysis of moving visual patterns. *Experimental Brain Research Supplementum 11*, 117–151.

Movshon, J. and W. Newsome (1996, December). Visual response properties of striate cortical neurons projecting to area MT in macaque monkeys. *Journal of Neuroscience 16*, 7733–7741.

MPEG-2 (1997). Information technology - Coding of moving pictures and associated audio information. ISO/IEC JTCI IS 13818-2 ISO standard, Geneva. www.iso.org.

MPEG-4 (1998). MPEG-4 video verification model version 8.0. ISO/IEC JTC1/SC29/WG11 ISO standard, Geneva. www.iso.org.

Murray, D. and B. Buxton (1987, March). Scene segmentation from visual motion using global optimization. *IEEE Transactions on Pattern Analysis and Machine Intelligence 9*(2), 220–228.

Newsome, W., K. Britten, and A. Movshon (1989). Neuronal correlates of a perceptual decision. *Nature 341*, 52–54.

O'Carroll, D., N. Bidwell, S. Laughlin, and E. Warrant (1996). Insect motion detectors matched to visual ecology. *Nature 382*, 63–66.

Pack, C. and R. Born (2001, February). Temporal dynamics of a neural solution to the aperture problem in visual area MT of macaque brain. *Nature 409*, 1040–1042.

Pack, C., J. Hunter, and R. Born (2005). Contrast dependence of suppressive influences in cortical area MT of alert macaque. *Journal of Neurophysiology 93*, 1809–1815.

Perrone, J. and A. Thiele (2001). Speed skills: Measuring the visual speed analyzing properties of primate MT neurons. *Nature Neuroscience 4*(5), 526–532.

Pesavento, A. and C. Koch (1999). Feature detection in analog VLSI. In *Proceedings of the 33rd Asiolmar Conference on Signals, Systems and Computers*, Monterey, CA.

Platt, J. (1989, July). *Constraint Methods for Neural Networks and Computer Graphics*. Ph.D. thesis, California Institute of Technology, Dept. of Computer Science.

Poggio, T. and V. Torre (1984, April). Ill-posed problems and regularization analysis in early vision. Technical Report 773, MIT AI Lab, Cambridge, MA.

Poggio, T., V. Torre, and C. Koch (1985, September). Computational vision and regularization theory. *Nature 317*(26), 314–319.

Pouget, A., P. Dayan, and R. Zemel (2003). Inference and computation with population codes. *Annual Reviews in Neuroscience 26*, 381–410.

Priebe, N., C. Cassanello, and S. Lisberger (2003, July). The neural representation of speed in macaque area MT/V5. *Journal of Neuroscience 23*(13), 5650–5661.

Priebe, N. and S. Lisberger (2004, February). Estimating target speed from the population response in visual area MT. *Journal of Neuroscience 24*, 1907–1916.

Rao, R. (2004). Bayesian computation in recurrent neural circuits. *Neural Computation 16*, 1–38.

Rao, R. and D. Ballard (1996). The visual cortex as a hierarchical predictor. Technical Report 96.4, Dept. of Computer Science, University of Rochester.

Rao, R. and D. Ballard (1999, January). Predictive coding in the visual cortex: A functional interpretation of some extra-classical receptive-field effects. *Nature Neuroscience 2*(1), 79–87.

Reppas, J., S. Niyogi, A. Dale, M. Sereno, and R. Tootell (1997, July). Representation of motion boundaries in retinotopic human visual cortical areas. *Nature 388*, 175–179.

Reyneri, L. (2003, January). Implementation issues of neuro-fuzzy hardware going toward HW/SW codesign. *IEEE Transactions on Neural Networks 14*(1), 176–194.

Rockland, K. and J. Lund (1983). Intrinsic laminar lattice connections in primate visual cortex. *Journal of Computational Neurology 216*, 303–318.

Salinas, E. and L. Abbott (1996). A model of multiplicative neural responses in parietal cortex. *Proceedings of the National Academy of Sciences USA 93*, 11956–11961.

Sarpeshkar, R., W. Bair, and C. Koch (1993). Visual motion computation in analog VLSI using pulses. In *Advances in Neural Information Processing Systems 5*, pp. 781–788. Cambridge, MA: MIT Press.

Sawaji, T., T. Sakai, H. Nagai, and T. Matsumoto (1998). A floating-gate MOS implementation of resistive fuse. *Neural Computation 10*(2), 486–498.

Schrater, P., D. Knill, and E. Simoncelli (2000, January). Mechanisms of visual motion detection. *Nature Neuroscience 3*(1), 64–68.

Sclar, G., J. Maunsell, and P. Lennie (1990). Coding of image contrast in central visual pathways of the macaque monkey. *Vision Research 30*(1), 1–10.

Shadlen, M., K. Britten, W. Newsome, and J. Movshon (1996, February). A computational analysis of the relationship between neuronal and behavioral responses to visual motion. *Journal of Neuroscience 16*(4), 1486–1510.

Simoncelli, E. (1993). *Distributed analysis and representation of visual motion.* Ph.D. thesis, MIT, Dept. of Electrical Engineering.

Simoncelli, E. and D. Heeger (1998). A model of neuronal responses in visual area MT. *Vision Research 38*(5), 743–761.

Simoncelli, E. P. (2003). Local analysis of visual motion. In L. M. Chalupa and J. S. Werner (Eds.), *The Visual Neurosciences*, pp. 1616–1623. Cambridge, MA: MIT Press.

Singh, A. (1991). *Optic Flow Computation: A Unified Perspective.* Los Alamitos, CA: IEEE Computer Society Press.

Smith, M., N. Majaj, and J. Movshon (2005, February). Dynamics of motion signaling by neurons in macaque area MT. *Nature Neuroscience 8*(2), 220–228.

Srinivasan, M., M. Lehrer, W. Kirchner, and S. Zhang (1991). Range perception through apparent image speed in freely flying honeybees. *Visual Neuroscience 6*, 519–535.

Stiller, C. and J. Konrad (1999, July). Estimating motion in image sequences. *IEEE Signal Processing 16*(4), 71–91.

Stocker, A. (2002, May). An improved 2-D optical flow sensor for motion segmentation. In *IEEE International Symposium on Circuits and Systems*, Phoenix, AZ, Volume 2, pp. 332–335.

Stocker, A. (2003, May). Compact integrated transconductance amplifier circuit for temporal differentiation. In *IEEE International Symposium on Circuits and Systems*, Vancouver, Volume 1, pp. 201–204.

Stocker, A. (2004, May). Analog VLSI focal-plane array with dynamic connections for the estimation of piece-wise smooth optical flow. *IEEE Transactions on Circuits and Systems-1: Regular Papers 51*(5), 963–973.

Stocker, A. and R. Douglas (1999). Computation of smooth optical flow in a feedback connected analog network. In M. Kearns, S. Solla, and D. Cohn (Eds.), *Advances in Neural Information Processing Systems 11*, pp. 706–712. Cambridge, MA: MIT Press.

Stocker, A. and R. Douglas (2004, May). Analog integrated 2-D optical flow sensor with programmable pixels. In *IEEE International Symposium on Circuits and Systems*, Vancouver, Volume 3, pp. 9–12.

Stocker, A. and E. Simoncelli (2005a). Constraining a Bayesian model of human visual speed perception. In L. K. Saul, Y. Weiss, and L. Bottou (Eds.), *Advances in Neural Information Processing Systems NIPS 17*, pp. 1361–1368. Cambridge, MA: MIT Press.

Stocker, A. and E. Simoncelli (2005b, March). Explicit cortical representation of probabilities is not necessary for optimal perceptual behavior. In *Computational Systems Neuroscience COSYNE abstract*, Salt Lake City.

Stocker, A. (2006, February). Analog integrated 2-D optical flow sensor. *Analog Integrated Circuits and Signal Processing 46*(2), 121–138.

Stone, L. and P. Thompson (1992). Human speed perception is contrast dependent. *Vision Research 32*(8), 1535–1549.

Stone, L., A. Watson, and J. Mulligan (1990). Effect of contrast on the perceived direction of a moving plaid. *Vision Research 30*(7), 1049–1067.

Sundareswaran, V. (1991, October). Egomotion from global flow field data. In *IEEE Workshop on Visual Motion*, pp. 140–145. Los Alamitos, CA: IEEE Computer Society Press.

Sze, S. (1981). *Physics of Semiconductor Devices* (2nd ed.). New York: John Wiley & Sons, Inc.

Szeliski, R. (1990, December). Bayesian modeling of uncertainty in low-level vision. *International Journal of Computer Vision 5*, 271–301.

Szeliski, R. (1996). Regularization in neural nets. In P. Smolensky, M. Mozer, and D. Rumelhart (Eds.), *Mathematical Perspectives on Neural Networks*, pp. 497–532. Mahwah, NJ: Lawrence Erlbaum Associates.

Tank, D. and J. Hopfield (1986, May). Simple "neural" optimization networks: An A/D converter, signal decision circuit and a linear programming circuit. *IEEE Transactions of Circuits and Systems 33*(5), 533–541.

Tanner, J. and C. Mead (1986). An integrated analog optical motion sensor. In S.-Y. Kung, R. Owen, and G. Nash (Eds.), *VLSI Signal Processing, 2*, pp. 59ff. Piscataway, NJ: IEEE Press.

Tautz, J., W. Zhang, J. Spaethe, A. Brockmann, A. Si, and M. Srinivasan (2004, July). Honeybee odometry: Performance in varying natural terrain. *PoLS Biology 2*(7), 0915–0923.

Thompson, P. (1982). Perceived rate of movement depends on contrast. *Vision Research 22*, 377–380.

Tikhonov, A. and V. Arsenin (1977). *Solutions of Ill-posed Problems*. Washington, DC: Winston and Sons, Scripta Technica.

Treue, S. and J. Maunsell (1996, August). Attentional modulation of visual motion processing in cortical areas MT and MST. *Nature 382*, 539–541.

Treue, S. and J. Trujillo (1999, June). Feature-based attention influences motion processing gain in macaque visual cortex. *Nature 399*, 575–579.

Turing, A. (1950, October). Computing machinery and intelligence. *MIND: A Quarterly Review of Psychology and Philosophy 236*, 59.

Ullman, S. (1979). *The Interpretation of Visual Motion*. Cambridge, MA: MIT Press.

Ullman, S. and A. Yuille (1987, November). Rigidity and smoothness of motion. Technical Report 989, MIT AI Lab, Cambridge, MA.

Van Santen, J. and G. Sperling (1984). Temporal covariance model of human motion perception. *Journal of the Optical Society of America A 1*, 451–473.

Verri, A. and T. Poggio (1989). Motion field and optical flow: Qualitative properties. *IEEE Transactions on Pattern Analysis and Machine Intelligence 11*(5), 490–498.

Vittoz, E. (1989). Analog VLSI implementation of neural networks. In *Proceedings of Journées d'Electronique et Artificial Neural Networks*, pp. 223–250. Lausanne; press politechnique.

von Neumann, J. (1982/1945). First draft of a report on the EDVAC. In B. Randall (Ed.), *The Origins of Digital Computers: Selected Papers*. Berlin: Springer.

Wallach, H. (1935). Über visuell wahrgenommmene Bewegungsrichtung. *Psychologische Forschung 20*, 325–384.

Watson, A. and A. Ahumada (1985). Model of human visual motion sensing. *Journal of the Optical Society of America A 2*, 322–342.

Weiss, Y. (1997). Smoothness in layers: Motion segmentation using nonparametric mixture estimation. In *Proceedings of IEEE Conference on Computer Vision and Pattern Recognition*, Puerto Rico, pp. 520–527.

Weiss, Y. and E. Adelson (1998, February). Slow and smooth: A Bayesian theory for the combination of local motion signals in human vision. Technical Report 1624, MIT AI Lab, Cambridge, MA.

Weiss, Y. and D. Fleet (2002). Velocity likelihoods in biological and machine vision *Probabilistic Models of the Brain*, pp. 77–96. In Bradford Cambridge, MA: Books, MIT Press.

Weiss, Y., E. Simoncelli, and E. Adelson (2002, June). Motion illusions as optimal percept. *Nature Neuroscience 5*(6), 598–604.

Wilson, G. and G. Pawley (1988). On the stability of the Traveling Salesman Problem algorithm of Hopfield and Tank. *Biological Cybernetics 58*, 63–70.

Wyatt, J. (1995). Little-known properties of resistive grids that are useful in analog vision chip designs. In C. Koch and H. Li (Eds.), *Vision Chips: Implementing Vision Algorithms with Analog VLSI Circuits*, pp. 72–89. Los Alamitos, CA: IEEE Computer Society Press.

Yamada, K. and M. Soga (2003, March). A compact integrated visual motion sensor for ITS applications. *IEEE Transactions on Intelligent Transportation Systems 4*(1), 35–42.

Yo, C. and H. Wilson (1992). Perceived direction of moving two-dimensional patterns depends on duration, contrast and eccentricity. *Vision Research 32*(1), 135–147.

Yu, P., S. Decker, H.-S. Lee, C. Sodini, and J. Wyatt (1992, April). CMOS resistive fuses for image smoothing and segmentation. *IEEE Journal of Solid State Circuits 27*(4), 545–553.

Yuille, A. (1989). Energy functions for early vision and analog networks. *Biological Cybernetics 61*, 115–123.

Yuille, A. and N. Grzywacz (1988, May). A computational theory for the perception of coherent visual motion. *Nature 333*(5), 71–74.

Yuille, A. and N. Grzywacz (1989). A mathematical analysis of the motion coherence theory. *International Journal of Computer Vision 3*, 155–175.

Index

Analog VLSI Circuits for the Perception of Visual Motion A. A. Stocker
© 2006 John Wiley & Sons, Ltd